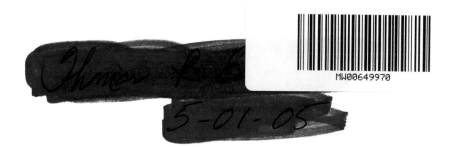

# Nondetects
# And
# Data
# Analysis

# STATISTICS IN PRACTICE

*Advisory Editor*

**Marian Scott**
*University of Glasgow, UK*

*Founding Editor*

**Vic Barnett**
*Nottingham Trent University, UK*

---

The texts in the series provide detailed coverage of statistical concepts, methods and worked case studies in specific fields of investigation and study.

With sound motivation and many worked practical examples, the books show in down-to-earth terms how to select and use an appropriate range of statistical techniques in a particular practical field. Readers are assumed to have a basic understanding of introductory statistics, enabling the authors to concentrate on those techniques of most importance in the discipline under discussion.

The books meet the need for statistical support required by professionals and research workers across a range of employment fields and research environments. Subject areas covered include medicine and pharmaceutics; industry, finance and commerce; public services; the earth and environmental sciences.

A complete list of titles in this series appears at the end of the volume.

# Nondetects
# And
# Data
# Analysis

## Statistics for Censored Environmental Data

**Dennis R. Helsel**
U.S. Geological Survey
Denver, CO

A JOHN WILEY & SONS, INC., PUBLICATION

Published by John Wiley & Sons, Inc., Hoboken, New Jersey.
Published simultaneously in Canada.

For general information on our other products and services please contact our Customer Care
Department within the U.S. at 877-762-2974, outside the U.S. at 317-572-3993 or fax 317-572-4002.

Wiley also publishes its books in a variety of electronic formats. Some content that appears in print,
however, may not be available in electronic format.

*Library of Congress Cataloging-in-Publication Data:*

Helsel, Dennis R.
    Nondetects and data analysis : statistics for censored environmental data / Dennis R. Helsel.
      p. cm. — (Statistics in practice)
    Includes bibliographical references and index.
    ISBN 0-471-67173-8 (cloth : acid-free paper)
      1. Environmental sciences—Statistical methods.  2. Survival analysis (Biometry).  I. Title
    II. Statistics in practice (Chichester, England)

QE45.S73H45 2005
363.73'01'1—dc22                                                     2004049165

Printed in the United States of America.

10 9 8 7 6 5 4 3 2 1

To my family

for their patience during what seemed to them

a never-ending process.

*Yesterday upon the stair*
*I saw a man who wasn't there*
*He wasn't there again today*
*I wish, I wish he'd go away.*

Hughes Mearns (1875–1965)

# Contents

# Preface

This book introduces methods for censored data, some simple and some more complex, to potential users who until now were not aware of their existence, or perhaps not aware of their utility. These methods are directly applicable to air quality, water quality, soils, and contaminants in biota, among other media. Most of the methods come from the field of survival analysis, where the primary variable being investigated is length of time. Here they are instead applied to environmental measures such as concentration. *Nondetects and Data Analysis* expands on ideas presented in my article, Less Than Obvious, published in 1990 in the journal *Environmental Science and Technology*. Several other authors have also suggested their use, beginning with Miesch in 1967. Yet these methods have not found their way into common use within any of the environmental sciences. It is my hope that this book will remedy that situation.

Within each chapter, examples have been provided in sufficient detail so that readers may apply these methods to their own work. Commercially available software was used so that methods would be easily available. Examples throughout the book were computed using Minitab® (version 14), one of several software packages providing routines for survival analysis. Other statistical software can also perform most or all of the methods used here. *

The website linked with this book:

**www.practicalstats.com/nada**

contains material for the reader that augments this textbook. Located on the website are: Answers to exercises computed using Minitab® as well as other software packages, Minitab® macros to perform methods not available in the commercial version, datasets used in this book, and as necessary, an errata sheet listing corrections to the text.

Comments and feedback on both the website and the book may be e-mailed to

nada@practicalstats.com

I sincerely hope that you find this book helpful in your work.

DENNIS HELSEL

---

* Mention of specific trade or product names does not imply endorsement by the U.S. Government.

# Acknowledgments

My sincere appreciation goes to a number of people who have made this book possible:

To the reviewers: Dr. Edward J. Gilroy, consultant; David K. Mueller, U.S. Geological Survey; and to a host of students who have reviewed portions of notes and drafts, making many suggestions and improvements.

To the scientists who have supplied their data to be used as real-life examples; to Stuart Giles, for the layout, design and typesetting; and finally to A. T. Miesch, who led the way decades ago.

# INTRODUCTION

# AN ACCIDENT WAITING TO HAPPEN

On January 28, 1986, the space shuttle *Challenger* exploded seventy-three seconds after liftoff from Kennedy Space Center, killing all seven astronauts on board and severely wounding the U.S. space program. In addition to career astronauts, on board was America's Teacher In Space, Christa McAuliffe, who was to tape and broadcast lessons designed to interest the next generation of children in America's space program. Her participation ensured that much of the country, including its school children, was watching.

What caused the accident? Would it happen again on a subsequent launch? Four months later the Presidential Commission investigating the accident issued its final report (Rogers Commission, 1986). It pinpointed the cause as a failure of O-rings to flex and seal in the 30° F temperatures at launch time. Rocket fuel exploded after escaping through an opening left by a failed O-ring. An on-camera experiment during the hearings by physicist Richard Feynman illustrated how a section of O-ring, when placed in a glass of ice water, failed to recover from being squeezed by pliers. The experiment's refreshing clarity contrasted sharply with days of inconclusive testimony by officials who debated what might have taken place.

The most disturbing part of the Commission's report was that the O-ring failure had been foreseen by engineers of the booster rockets' manufacturer, who were unable to persuade managers to delay the launch. Rocket tests had previously shown evidence of thermal stress in O-rings when temperatures were 65° F and colder. No data were available for the extremely low temperatures predicted for launch time. Faxes sent to NASA on January 27th, the night before launch, presented a graph of damage incidents to one or more rocket O-rings as a function of temperature (Figure 1). This evidence given in the figure seemed inconclusive to managers – there were few data and no apparent pattern.

The Rogers Commission noted in its report that the above graph had one major flaw – flights where damage had not been detected were deleted. The Commission produced a modified graph, their assessment of what should have been (but was not) sent to NASA managers. Their graph added back in the non-detected values (Figure 2). By including all recorded data, the Commission proved that the pattern was a bit more striking.

What type of graph could the engineers have used to best illustrate the risk they believed was present? The vast store of information in nondetects is contained in the proportions at which they occur. A simple bar chart could have focused on the proportion of O-rings exhibiting damage. For a possible total of 3 damage incidents in each rocket, a graph of the proportion of failure incidents by ranges of 5 degrees in temperature is shown in Figure 3. The increase in the proportion of damaged O-rings with lower temperatures is clear.

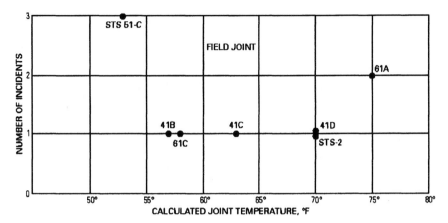

**Figure 1.** Plot of flights with incidents of O-ring thermal distress – "nondetects" deleted. (Figure 6 from Rogers Commission, 1986, p. 146)

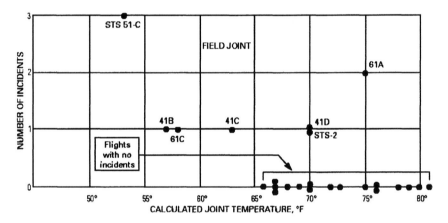

**Figure 2.** Plot of flights with and without incidents of O-ring thermal distress – "nondetects" included. (Figure 7 from Rogers Commission, 1986, p. 146)

In Figure 1, the information content of data below a (damage) detection threshold was discounted, and the data ignored. Not recognizing and recovering this information was a serious error by engineers. Today the same types of errors are being made by numerous environmental scientists. Deleting nondetects, concentrations below a measurement threshold, obscures the information in graphs and numerical summaries. Statements such as the one below from the ASTM committee on intra-laboratory quality control are all too-common:

> Results reported as "less than" or "below the criterion of detection" are virtually useless for either estimating outfall and tributary loadings or concentrations for example. (ASTM D4210, 1983)

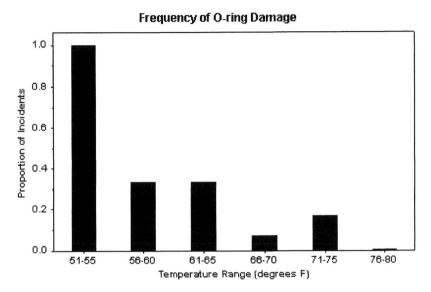

**Figure 3.** O-ring thermal distress data, re-expressed as proportions.

A second, equally serious error occurred prior to the *Challenger* launch, when managers assumed that they possessed more information on launch safety than was contained in their data. They decided to launch without knowing the consequences of very low temperatures. According to Richard Feynman, their attitude had become "a kind of Russian roulette....We can lower our standards a little bit because we got away with it the last time" (Rogers Commission, 1985, p.148). A similar error is now frequently made by environmental programs that fabricate numbers, such as one-half the detection limit, to replace nondetects. Substituting a constant value is even mandated by some Federal agencies – it seemed to work the last time they used it. Its primary error lies in assuming that the scientist/regulator knows more information than what is actually contained in their data. This can easily result in the wrong conclusion, such as declaring that an area is 'clean' when it really is not. For the *Challenger* accident, the consequences were a tragic one-time loss of life. For environmental sciences, the consequences are likely to be more chronic and continuous. The health effects of many environmental contaminants occur in the same ranges as current detection limits. Assuming that measurements are at one value when they could be at another is not a safe practice, and as we shall see, totally unnecessary. Fabricating numbers for concentrations could also lead to unnecessary expenditures for cleanup, declaring an area is worse than it actually is. With the large (but limited) amounts of funding now spent on environmental measurements and evaluations, it is incumbent on scientists to use the best available methodologies. In regards to deleting nondetects, or fabricating numbers for them, there are better ways.

When interpreting data that include values below a detection threshold, keep in mind three principles:

1. Never delete nondetects;
2. Capture the information in the proportions; and
3. Never assume that you know more than you do.

This book is about what else is possible.

# 1

# In Focus

They could be many things – trace metal concentrations in the bodies of mayflies in pristine streams and streams with industrial outfalls; particulates in the atmosphere inside and outside of a national park; cadmium concentrations in soils upwind and downwind of an old smelter site; blood lead levels in children from cities and children from rural areas. Comparing two groups of data, one a possibly contaminated test group and the other a control group, is a basic design in environmental science. Are concentrations in the test group higher than in the control group?

The classic approach for this design is the two-sample t-test. If data distributions do not follow a normal distribution, the nonparametric Mann-Whitney (also called Wilcoxon rank-sum) test is used instead. With either test, a roadblock looms in the data shown in Table 1.1 – there are values below detection limits; several detection limits.

The most common method in environmental studies for dealing with such data is to substitute one-half the detection limit. Many scientists have used this method over the years, from older studies (Nehls and Akland, 1973) to current research (Harris and others, 2003). Procedures manuals for at least two Federal agencies recommend the practice (Environmental Protection Agency, 1998; Army Corps of Engineers, 1998). Substitution for the Table 1.1 data produces the data of Table 1.2, and a Mann-Whitney test p-value of 0.015. The equivalence of the groups is rejected, and the test group is declared higher than the control group. Expensive remediation actions might be mandated for conditions that have caused the elevated concentrations in the test group. Soil is removed. Industrial equipment is modified. Wells are abandoned. People are given new medications.

Now let's pull back a curtain. These data were not field data, but were computer generated. By generating data, the true situation is known. All of the data in Table 1.1 came from the same distribution – there is actually NO difference in their mean or median levels (see Figure 1.1). For the original uncensored data, the Mann-Whitney test produced a one-sided p-value of 0.43, stating that there is no evidence for difference between the two groups. Any reasonable method for analyzing the data with nondetects (statisticians call these "censored data") should also find no difference in the two groups. For example, in chapter 9 a Wilcoxon score test is presented, a nonparametric test to compare two groups of data with multiple thresholds. No substitution is involved, and the test produces a p-value of 0.47 for the censored Table 1.1 data. No difference. No contamination. No remediation.

The only conclusion possible is that the most commonly used method in environmental studies today, substitution of one-half the detection limit, is NOT a reasonable method for interpreting censored data.

By substituting values of one-half the detection limits and running a simple test,

5

**Table 1.1.** Contaminant concentrations with multiple detection limits in a test and a control group.

| Control Group | | Test Group | |
|---|---|---|---|
| <1 | <1 | <2 | <5 |
| <1 | <1 | <2 | <5 |
| <1 | <1 | 3.3 | <5 |
| <1 | 4.1 | 3.4 | <5 |
| 1.0 | 7.0 | <2 | 4.7 |
| 1.8 | 7.5 | 12.2 | <5 |
| 2.2 | 15.4 | <5 | 22.5 |
| <2 | | 6.6 | |

**Table 1.2.** Contaminant concentrations in a test and a control group after substituting one-half the detection limit for nondetects.

| Control Group | | Test Group | |
|---|---|---|---|
| 0.5 | 0.5 | 1.0 | 2.5 |
| 0.5 | 0.5 | 1.0 | 2.5 |
| 0.5 | 0.5 | 3.3 | 2.5 |
| 0.5 | 4.1 | 3.4 | 2.5 |
| 1.0 | 7.0 | 1.0 | 4.7 |
| 1.8 | 7.5 | 12.2 | 2.5 |
| 2.2 | 15.4 | 2.5 | 22.5 |
| 1.0 | | 6.6 | |

expensive cleanup measures may be implemented where none are needed. The fundamental problem with substituting one-half, or any other function of the detection limit, is in the statement that something is known (the values for nondetects) that really is not known. The test procedure takes our word that the value is actually 0.5 times the detection limit, not some other value below the limit. The true value may have been anywhere below the detection limit, as far as we know. To compound this problem, the value substituted is not a function of anything known about the media sampled (the organism, water, or soil). It is a function of the precision, or lack of it, in the laboratory. It may be a function of the process used by that laboratory, and laboratories differ considerably in how they compute detection limits (see Chapter 3). It may be a function of time, or of the dilution of the samples, or of interferences in some samples but not others, or of other conditions in the laboratory process. As in this example, studies using substitution can easily impose an artificial signal that originally was not there.

An error can also be made in the opposite direction when using substitution. It is trivial to generate data whose levels originally do differ between groups, but after censoring and substitution a t-test or Mann-Whitney test fails to find any evidence of difference. The inability to see differences again results from specifying artificial

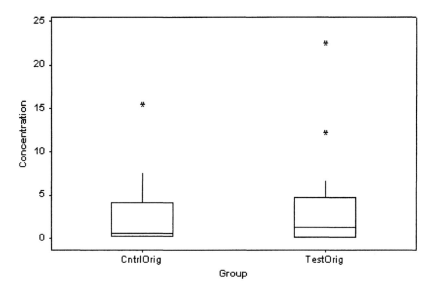

**Figure 1.1.** Boxplots for data of Table 1.1 prior to setting artificial detection limits. Mann-Whitney test p-value (uncensored data) = 0.43.


values when the true values are unknown. The result is not just an incorrect conclusion by a hypothesis test. In the real world, contamination goes unnoticed. Remediation goes undone. Public health is unknowingly threatened.

There are better ways.

# 2

# THE THREE APPROACHES

Measurements whose values are known only to be above or below a threshold are called "censored" data in the statistical literature. Censored data have been an integral part of the social sciences, economics, and medical and industrial statistics for some time. Procedures developed in those fields allow censored data to be incorporated into the computations of summary statistics, regression, and hypothesis tests. Yet these procedures are rarely used in environmental studies, where censored data are commonly encountered as values below a detection limit. Called "less-thans" or "nondetects" (shown as the white bars in Figure 2.1), these low values for contaminants such as trace metals or organic compounds are known inexactly. Because low values are usually plotted to the left on a graph, nondetects are labeled as "left-censored", with values lying somewhere to the left of the detection threshold. Instead of having an exact measure of concentration or mass shown as the tip of the shaded bars for known concentrations, censored data in Figure 2.1 lie somewhere between 0 and the tip of the white bar, somewhere along its length. But the exact concentration is unknown.

Nondetects are an issue in several disciplines, and within many sub-disciplines within the environmental sciences. They are encountered in air quality (Rao et al., 1991), marine studies (Huybrechts et al., 2002), streamflow analyses (Kroll and Stedinger, 1996), precipitation chemistry (Ahn, 1998), groundwater pump tests (Wen, 1994), human exposure to toxins (Perkins et al., 1990; Vance et al., 1988), human HIV and other viral studies (Lynn, 2001), contaminants in the bodies of a variety of animals (Harris et al., 2003; Hobbs et al., 2003), sediment chemistry (Clarke, 1998), astronomy (Isobe et al., 1996), rock and mineral chemistry (Miesch, 1967), and in studies of stream and ground water quality (Helsel and Gilliom, 1986). The issue of how best to deal with left-censored data is obviously of concern to many scientists.

Data in medical and industrial studies are most often right-censored, known only to be greater than a threshold value (Lee and Wang, 2003), as shown in Figure 2.2. One example might be to measure the length of time people live after receiving medical treatment. In Figure 2.2 the left side of the bars represents the time at which people are diagnosed, and treatment begins. This occurs at different times for different patients. Two treatments might be compared for their effectiveness, measured as the median survival time after treatment. The preferred treatment is the one resulting in longer life expectancies. For patients still living after the experiment concludes, represented by the far right end of the plot in Figure 2.2, survival times are not known exactly. Their times are "greater than" values, shown in Figure 2.2 as white bars.

Few papers in environmental science have recommended, or actually used, cen-

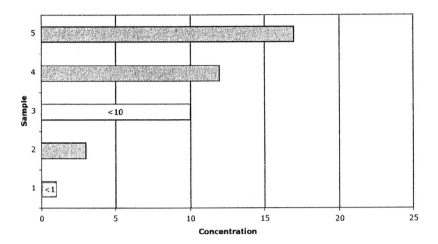

**Figure 2.1.** Five observations, including two observations recorded as below one of the two detection limits. Left-censored data.

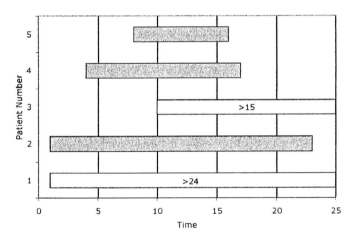

**Figure 2.2.** Five observations, including two patients surviving past the end of the study (unshaded bars). Right-censored data.

sored data techniques adopted from other disciplines. Perhaps the first was Miesch (1967), who promoted the use of maximum likelihood estimation to compute mean values of mineral resource data. Millard and Deverel (1988) demonstrated how score tests, two-group tests for censored data, could be used for water quality studies. Helsel (1990) advocated the use of survival analysis techniques for hypothesis testing and regression. Slyman and others (1994) used maximum likelihood for regression of censored concentration data. She (1997) used Kaplan-Meier estimates of summary statistics to describe water-quality data. A guide to the use of censored data techniques for environmental studies was published by Akritas (1994) as a chapter in volume 12 of the *Handbook of Statistics*. However, neither these examples nor the *Handbook* chapter have changed standard practices. None of these recommended methods are in regular use today in the environmental sciences.

Instead, two overly simplistic methods are commonly used when censored data are encountered. The first is to delete censored values. Deleting the lowest values obviously produces biased results. The statistics or tests which result do not apply to the entire data set collected, but only to a part of the data on the high end of the distribution. The argument for deletion is usually that the only interest is in detected observations (Kolpin et al., 2002a; Lundgren and Lopes, 1999) – that is, "when it is present, what are the concentrations?". However, "present" is an inexact criteria – as technology changes and detection limits lower, lower concentrations are "present" that would not have been before. Comparisons among groups of data within and between studies will depend on the detection limits in force at the time each was analyzed. Results may differ had a different detection limit been in force if only detected observations are considered. Statistics using only detected observations overstate the levels actually encountered, and are difficult to put into context for the entire distribution. This leaves the authors vulnerable to claims of bias (Till, 2003). For example, comparisons using only detected observations might contrast the mean or median of the top 30% of one group with the top 60% of another. Such a procedure does not represent information about the two original distributions, and indicates nothing about whether the two distributions, detected and nondetected observations taken together, can be distinguished. The frustrating part, of course, is that deleting nondetects is never necessary. There are better ways.

The second method commonly used for dealing with censored data is to assign an arbitrary fraction of the detection limit to each censored observation ("substitution", or more honestly, fabrication). In numerous studies over the years, one-half the detection limit has been substituted for censored values, from Nehls and Akland (1973) to Buckley et al. (1997), and to Hobbs et al. (2003). Chapter 1 pointed out the dangers in doing so. Substitution can induce a signal not present in the original data, or obscure a signal that was actually there. Using simulations to compare differing methods of computing summary statistics for censored data, Singh and Nocerino (2002) report that "substitution [by one-half dl] resulted in a biased estimate of the mean with the highest variability...". Their work was for the simplest case of data with one detection limit; errors by substitution methods increase when multiple detection limits are present.

The lack of care in using censored data is evidence of a general skepticism about the information content of these observations. In truth, a great deal of information is available in censored data. If efficient methods are used, the information extracted from them is almost equal to that for data with single known values. Their information is primarily contained in the proportion of data below the threshold value(s). Knowing that for one data set, three quarters of the data are below the detection limit, while a second data set has only ten percent below the same limit, strongly indicates that the first data set contains lower values than the second. This is evident without any knowledge of values above the detection limit. Efficient procedures for censored data combine the values above the detection limit(s) with the information contained in the proportion of data below the detection limit(s).

***Example – Information Content of Nondetects:***
Estimate the center (median) of the following data.

<1    <1    <1    <1    <1    <1    <1    5    12    22

If there were no information content in nondetects, the seven censored values could be discarded and the median of the three remaining values would equal 12. However, 12 is a very poor estimate of the center of a data set in which 70% of the observations are below 1. A much better estimate would be <1. There is a great deal of information in the lowest values in the data set – the issue is how to best extract that information.

There are three approaches for extracting information from datasets that include nondetects.

1. Substitution  –  fabricating numbers.
2. Maximum likelihood estimation  –  procedures assuming a specific distribution.
3. Nonparametric Methods  –  distribution-free procedures.

The first approach, substitution, is best avoided. It is discussed in this text only because it is so common, and so inadequate. The other two methods are viable approaches for analysis of censored data, and are used as standard methods in disciplines other than environmental science. A general description of each approach follows.


**Approach 1: Substitution**

Substitution methods compute statistics for censored data by fabricating values as functions of the recorded detection limit for every nondetect. Summary statistics, hypothesis tests, and regression models are then calculated using these fabricated numbers, along with measured values above detection limits. Substitution methods are widely used, but have no theoretical basis. Numerous papers have shown that substitution methods do not work well in comparison to the other two types of procedures (for example, see Gleit, 1985; Helsel and Cohn, 1988; Singh and Nocerino, 2002). Substitution is used most often because it is easy. No special software is required. However, fabrication of values is not defensible – the choice of the value to substitute is arbitrary, and statistical results usually differ depending on the value chosen.

The situation in which substitution fails most miserably is when there are multiple detection limits. The values substituted depend on the conditions which determined the detection limit, such as lab precision or sample matrix interferences (see Chapter 3). Values substituted do not necessarily bear any relation to the true value in the sample. For multiple detection limits, either MLE or nonparametric methods will far outperform substitution methods.

## Approach 2: Maximum likelihood estimation

Maximum likelihood estimation (MLE) has been used sporadically in environmental studies — Owen and DeRouen (1980) for air quality and Miesch (1967) for geochemistry are two early examples. MLE uses three pieces of information to perform computations: a) numerical values above detection limits, b) the proportion of data below each detection limit, and c) the mathematical formula for an assumed distribution. Data both below and above the detection limit are assumed to follow a distribution such as the lognormal. Parameters are computed that best match a fitted distribution to the observed values above each detection limit and to the percentage of data below each limit.

The most crucial consideration for MLE is how well data fit the assumed distribution. A major problem with MLE is that for small data sets there is often insufficient information to determine whether the assumed distribution is correct or not, or to estimate parameters reliably. MLE has been shown to perform poorly for data sets with less than 25 to 50 observations (Gleit, 1985; Shumway et al., 2002). For larger data sets, MLE is an efficient way to estimate parameters, given that the chosen distribution is correct. The term "efficient" means that the fitted parameters have relatively small variability, so that their confidence limits are as small as possible. For data sets of at least 50 observations, and where either the percent censoring is small (so that the distributional shape can be evaluated) or the distribution can be assumed from knowledge outside the data set, MLE methods are the method of choice.

MLE methods are computed by solving a likelihood function L, where for a distribution with two parameters $\beta_1$ (mean) and $\beta_2$ (variance), $L(\beta_1, \beta_2)$ defines the likelihood of matching the observed distribution of data. The function L increases as the fit between the estimated distribution and the observed data improves. The parameters $\beta_1$ and $\beta_2$ are varied in an optimization routine, choosing values to maximize L. In practice it is the natural logarithm ln(L) rather than L itself that is maximized, where ln(L) is the "log-likelihood", often (though not necessarily) a negative number. Maximizing ln(L) is accomplished by setting the partial derivatives of ln(L) with respect to the two parameters equal to zero.

$$\frac{d\left(\ln L[\beta_1]\right)}{d(\beta_1)} = 0 \quad \text{and} \quad \frac{d\left(\ln L[\beta_2]\right)}{d(\beta_2)} = 0 \qquad (2.1)$$

The exact equation for L will change depending on the assumed distribution and the process under study (estimation of a mean, linear regression, etc.). However, in each case the likelihood function L is the product of two component pieces, one for censored observations and one for uncensored (detected) observations. In the uncensored piece is the probability density function p[x], the equation describing the frequency of observing individual values of x. In the censored piece is the survival function S[x], which is the probability of exceeding the value x. S[x] equals 1-F[x], where F[x] is the cumulative distribution function (cdf) of the distribution, the probability of being less than or equal to x. Either S[x] or F[x] can be used when writing the likelihood function.

In the most general case, L can be considered to be the product of three pieces, where the censored data component is split into two, one for left-censored and one for right-censored data:

$$L = \prod p[x] \prod (F[x]) \prod S[x] \tag{2.2}$$

where p[x] is the pdf as estimated from the detected observations, (F[x]) is the cdf as determined by left-censored observations, and S[x] is the survival function as determined by right-censored observations ("greater-thans"). Greater-thans are not typically found among environmental data, and so likelihood functions in environmental studies typically deal with only the first two pieces.

For censored data, two variables x and $\delta$ are required to represent each observation. The value for the measurement, or for the detection limit, is given by x. The indicator variable $\delta$ is a 0/1 variable that designates whether an observation is censored (0) or detected (1). As one of the simpler likelihood functions, the equation for L when estimating the mean and standard deviation of a normal distribution using MLE is:

$$L = \prod p[x_i]^{\delta_i} \bullet F[x_i]^{1-\delta_i} \tag{2.3}$$

where $\delta$ is as defined above, and for a normal distribution the pdf is

$$p[x] = \frac{\exp\left[-\frac{1}{2}\left(\frac{x-\mu}{\sigma}\right)^2\right]}{\sigma\sqrt{2\pi}} \tag{2.4}$$

For detected observations $\delta = 1$ and the second term in equation 2.3 becomes 1 and so drops out. For censored observations, $\delta = 0$ and the first term becomes 1 and so drops out. The cumulative distribution function for a normal distribution is

$$F[x] = \Phi\left[\frac{x-\mu}{\sigma}\right] \tag{2.5}$$

where $\Phi$ is the cdf of the standard normal distribution

$$\Phi[y] = \frac{1}{\sqrt{2\pi}} \int_0^y \exp\left(-u^2/2\right) du \tag{2.6}$$

After substituting in the above and setting the partial derivatives of ln(L) equal to 0 (equation 2.1), the nonlinear equations are solved by iterative approximation using the Newton-Raphson method. The solution provides the parameters mean and standard deviation for the distribution that best matches both the pdf and cumulative distribution function (or 1 − survival function) estimated from the data. In other words, the estimates of mean and standard deviation will be the parameters for the assumed distributional shape that had the highest likelihood of producing the observed values for the detected observations and the observed proportion of data below each of the detection limits.

Likelihood methods can be used when performing an hypothesis test. The test is set up to determine whether $\beta = 0$, where $\beta$ is the parameter of interest. This could be a slope coefficient in regression, or an estimate of the difference between two population means. The null hypothesis of $\beta = 0$ is compared to an alternative that $\beta \neq 0$ using one of two types of test procedures, either likelihood-ratio tests or Wald's tests.

Likelihood-ratio tests are based on the value for the log of the likelihood function, ln(L). The test compares the log-likelihoods for two models, one where $\beta$ = the value chosen by MLE, and the second for the "null" state, $\beta = 0$. The test statistic takes the form of [2 lnL($\beta$) − (2 lnL(0)], resulting in a large positive value if $\beta \neq 0$. This difference is the likelihood-ratio test statistic, and is compared to a chi-squared distribution to produce the p-value for the test. Likelihood-ratio tests are the form used by most statistical software that perform maximum likelihood.

Wald's test statistics take a form similar to t-tests in regression. The numerator of the test statistic is the MLE value for the coefficient $\beta$, and the denominator is the standard error of $\beta$. Their ratio is compared to a standard normal distribution. Wald's tests are generally not considered as accurate as are likelihood-ratio tests and the latter are preferred, though the differences in p-values are often small.

## Approach 3: Nonparametric methods

Nonparametric methods are so named because they do not involve computing "parameters", such as the mean or standard deviation, of an assumed distribution. Instead they use the relative positions (ranks) of data, a reflection of the data's percentiles. Because these methods do not require an assumption about the distribution of data, they are sometimes called "distribution-free" methods. Nonparametric methods are especially useful for censored data because they efficiently use the available information. Nondetects are known to be lower than values above their detection limit, and so are ranked lower. These methods do not require estimates of the unknown distances between nondetects and detected values, but only their relative order.

Nonparametric methods are now commonly used in the environmental sciences. There is general recognition that many variables measured in natural systems have skewed distributions, and normal theory tests work well only after transformations. Textbooks such as Gilbert (1987) and Helsel and Hirsch (2002) have demonstrated nonparametric procedures and their usefulness to environmental studies. However, nonparametric score tests, developed for data with multiple thresholds, are still not familiar to most environmental scientists, and are woefully underutilized. Score tests are extensions of the more familiar rank-sum, sign, and contingency table tests to situations with multiple thresholds. They are found in statistical software along with other methods for survival analysis.

### Application of survival analysis methods to environmental data

Consider a typical survival analysis problem, a test of whether light bulbs with a new filament composition last longer than those with the existing filament. A group of 15 light bulbs for each filament type is connected to power, and the length of time each burns is measured. After 48 hours, it is decided that this sample size is too small, and 20 additional light bulbs of each type are added to the test. After six weeks (1008 hours) from when it was begun, the experiment is stopped. By that time, many of the bulbs have burnt out, and their burn lengths recorded. However, some of the bulbs started in the second batch are still burning when the experiment ends. Their burn lengths are recorded as "greater than 960 hours", because they were still burning after 1008 – 48 hours of use. A few of the bulbs in the original batch are also still burning, and their lengths are recorded as "greater than 1008 hours".

Of interest is whether bulbs of both filament types have the same mean or median burn length. If all bulbs had burnt out, the lengths for every bulb would be known and a t-test or rank-sum test could be used to test for differences. However, for some of the bulbs the actual length is not known, but are censored as "greater-thans". Because there are two different thresholds resulting from the different times a bulb entered into the experiment, these data sets are also "multiply-censored". Survival analysis methods were designed for right-censored data sets with multiple censoring thresholds.

Environmental data are also often censored, with a number of nondetect values included in the data set. They are often multiply-censored, as detection limit thresholds change over time or with varying sample characteristics or among different laboratories. The primary difference between environmental and industrial/medical data is that environmental data are dominantly left-censored. Uncertainty occurs at the lower end, for values typically plot on the left side of a graph. Though some survival analysis software can use left-censored data, many are hard-wired for right-censored data. Left-censored data must be transformed into right-censored data before these routines can be used.

To demonstrate this transformation, consider a left-censored data set with 5 observations previously shown in Figure 2.1:

<1,    3,    <10,    12    17

Data measured when the detection limit was 1 are the values of <1, 3, and 12. Data measured when the detection limit was 10 are the values of <10 and 17. In Figure 2.1 these data were plotted as a bar chart. The nondetects were shown as open bars, and the detected values as shaded bars.

Figure 2.3 shows these same data, along with the dark bars drawn down from the upper end of the plot in addition to those from the lower end. The upper end is set at 25, so the lengths of the new dark bars are 8, 13 and 22 corresponding to the detected values of 17, 12 and 3 (25–17, 25–12, and 25–3). The lengths of the two dark bars for the censored values are >15 and >24.

The bars drawn down from the top represent the same data, on an alternate measurement scale. These new bars, which include greater-thans, are right-censored data. Left-censored data (the gray and white bars) can be transformed to right-censored data (the dark bars) by subtracting each value from the same large constant (equation 2.7), in this case 25. The constant can be any value larger than the maximum in the original data. This left-to-right transformation is called "flipping" the data distribution and is a linear transformation – the transformation does not alter the shape of the data distribution other than to reverse its direction. If the survival analysis software available to you is designed only for right-censored data, environmental data can be 'flipped' into a right-censored form, and the data analyzed.

$$\text{Flipped Data} = \text{Constant} - \text{Original Data} \qquad (2.7)$$

For example, SAS's PROC LIFETEST performs nonparametric tests of differences between groups of survival data. The procedure handles only right-censored

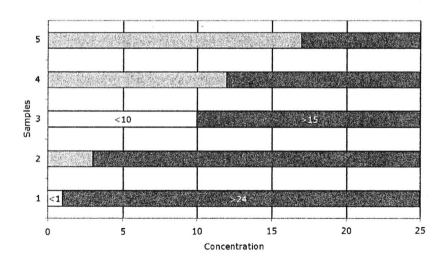

**Figure 2.3.** Five observations, including two observations recorded as below one of the two detection limits. Bars from the top down show conversion to right-censored data.

"greater thans". All of *Statistica*'s survival routines work only with right-censored data. Data with nondetects must first be flipped in order to use these routines. Some MLE software allows left-censored data to be analyzed directly, without the need for a transformation. Minitab's® MLE routines and SAS's® PROC LIFEREG perform censored regression using maximum likelihood where left-censored data can be coded directly into the model statement. In general, nonparametric methods for survival analysis are not coded to handle left-censored data. MLE software may or may not be.

Nonparametric hypothesis tests on flipped data will have the same test results for determining significance, the same p-values, as if the software had allowed the original left-censored data to be used as input. The p-values can be directly used to compare differences between distributions, etc. Parametric methods will also produce the same results for right and left censoring as long as care is taken to code in the lower bound of zero present in almost all environmental data (see Chapter 4). Data are transformed by flipping only to accommodate the software input requirements. Measures of location (mean, median and other percentiles) using flipped data must go through a reverse transformation to obtain estimates in the original units. Measures of spread or variability, such as the standard deviation and IQR, are the same for both original and flipped data, requiring no retransformation. Slopes of regression equations using flipped response (Y) variables must have their signs reversed to represent slopes of the original data.

### Power transformations with survival analysis

Parametric survival analysis methods, those computed using maximum likelihood, require that data follow a specific distribution. If after looking at plots or normality tests the data do not appear to follow this distribution, prior transformation of the data is necessary. Flipping the data from left- to right-censored does nothing to alter the skewness or outliers of a data set. Power transformations such as the square root or logarithm that change the shape of a distribution must be done prior to flipping the data into a right-censored format.

The order of processing is therefore:

1) Decide whether a power transform is necessary to alter the data's shape to be closer to the assumed distribution. If so, transform the data.
2) Flip the transformed data to produce a right-censored distribution (equation 2.7).
3) Compute the survival analysis test and interpret the test results.
4) Convert estimates of mean or median back to left-censored format by subtracting from the constant used to flip the data. If necessary, retransform using the reverse power transformation to get parameter estimates in original units.

**Parallels To uncensored methods**

A list of maximum likelihood and nonparametric methods that can be used to analyze censored data is given in Table 2.1. On the left are familiar methods used for uncensored data sets, data without nondetects. On the right are the equivalent methods used for censored data. Some of the censored methods are direct extensions of the uncensored procedure. Others are computationally very different, but perform the same function.

**Table 2.1.** Parallel Statistical Methods for Uncensored and Censored Data.

| Methods for Uncensored Data Sets | Methods for Censored Data Sets |
|---|---|
| Computing Summary Statistics || 
| Descriptive Statistics | Kaplan-Meier, MLE or ROS estimates |
| Comparing Two Groups ||
| t-test | Censored regression with 0/1 group indicator |
| Wilcoxon rank-sum test | Generalized Wilcoxon test |
| paired t-test | Censored CI on differences |
| (paired) sign or signed-rank test | PPW or Akritas tests |
| Comparing Three or More Groups ||
| ANOVA | Censored regression with 0/1 group indicators |
| Kruskal – Wallis test | Generalized Wilcoxon test |
| Correlation ||
| Pearson's r | Likelihood r |
| Kendall's tau | Kendall's tau-b |
| Linear Regression ||
| Regression | Censored regression |
|  | Logistic regression |
| Robust Regression | Proportional hazards (Cox) regression |
| Theil-Sen median line | Akritas-Theil-Sen median line |

# 3

# REPORTING LIMITS

"Reporting limit" is an intentionally generalized term that represents a variety of thresholds used to censor analytical results. It is a limit above which data values are reported without qualification by the analytical laboratory. If the terms "detection limit" and "nondetect" were not so ubiquitous, most of the occurrences of the term "detection limit" in this book would be replaced by "reporting limit". It would be best to reserve the term "detection limit" for its more specific meaning described in the discussion below. However, it is pointless to fight the battle for changing what has become so commonplace. Therefore, outside of this chapter I use the term "detection limit" in its most generic sense – as a reporting limit. But inside this chapter, the difference between detection limits and other types of reporting limits is an important one to understand.

Reporting limits are set in a variety of ways, and for a variety of purposes. As stated in a report summarizing the calculation of reporting limits (USEPA, 2003): "one conclusion that can be drawn is that detection limits are somewhat variable and not easy to define". Yet there are several things each type of reporting limit has in common. Each is a threshold computed so that measured values falling below that threshold are treated differently than those falling above. Most reporting limits are based on a measure of the variability or noise inherent in the laboratory process. The two general classes of reporting limits are split between those that assume this noise is constant over different concentrations, versus those that model the noise as a function of concentration. The standard deviation of repeated measurements is used to quantify the noise of the analytical process. First we discuss reporting limits based on constant standard deviation, and then those based on varying standard deviation.

From the data users' point of view, any method that changes a single numerical measurement into a censored value prior to reporting the data to the user is a 'reporting limit'. It may have been developed in a variety of ways, but all require the user to somehow interpret data labeled as a 'nondetect' or 'less-than'. The focus of this book is to provide methods that deal with data censored at reporting limits, regardless of the type of limit employed. However, knowledge by the data user of the type of limit employed can lead to better data analysis.

## Limits when the standard deviation is considered constant

Reporting limits are computed with a single value for the standard deviation when it is assumed that the noise of the measurement process is constant from concentrations of zero up to the highest reporting limit. Reporting limits for constant standard deviation can be classified into two general types, most often called detection limits and quantitation limits. The two types of limits differ in how they are computed, and in what they represent. These are the most commonly-used names for each type of limit, and generally describe the functions each type of limit serves. They or variants of them are the names used by the U.S. Environmental Protection Agency (USEPA) and by most laboratories following USEPA procedures. Another school of practice uses names as defined by Currie (1968) — decision limits and detection limits, respectively. Both schools have stated that it is "unfortunate" that the other uses the names that they do. For the data user it is unfortunate that there are multiple schools of names! The terms used here were chosen because they were the most prevalent. As time goes on, some standardization will hopefully be achieved.

### The detection limit

Values measured above this threshold are unlikely to result from a true concentration of zero.

A detection limit is a threshold below which measured values are not considered significantly different from a blank signal, at a specified level of probability. Measurements above the detection limit evidence a nonzero signal (at a given probability), indicating that the analyte is present in the sample. Other terms used for this type of threshold have included the "critical value" and "decision level" of Currie (1968), as well as the "method detection limit" or MDL (USEPA, 1982) and the "limit of detection" or LOD (Keith, 1992). USEPA (2003) discusses the differences among these variations of what is generally recognized as thresholds having the same objective – to distinguish samples with a concentration signal from those without a signal.

The first step in computing a detection limit is to estimate the inherent variation to be expected at a concentration of zero, where no analyte is present. Currie (1968) envisioned repeated measurements of blank solutions, but this is difficult to do successfully. With the MDL definition this variation is approximated by repeated measurements of a standard solution of low concentration. Figure 3.1 illustrates the process, using a standard at a concentration value of 2. The analyst is assuming that the measurement error at zero concentration is the same as at the low standard concentration – the standard deviation is constant between zero and the concentration standard. Measurement variation is almost always assumed to follow a normal distribution around the true value. The left-hand curve in Figure 3.1 illustrates the possible measured concentrations when the true concentration is zero. One-half of the measurements would be negative. The y-axis in the figure is the number of measurements for each value of concentration. Though the most frequently observed value is the true concentration at the center of each normal curve (assuming 100% recovery and no other bias), a variety of other measurements result from the same sample, due to random variation in the measurement process. The detection limit is set near the upper end of the distribution centered on zero (Figure 3.2). When an instrument measures a value above this limit, the concentration is unlikely to have

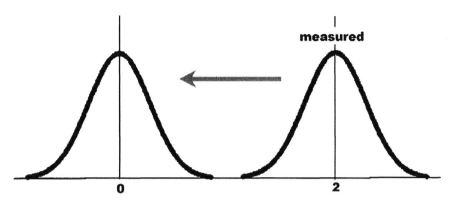

**Figure 3.1.** Error distribution of a measured standard at a concentration of 2, which is then imputed to also apply to a true zero concentration.

resulted from a true concentration of zero – unlikely to be a false positive.

The curve in Figure 3.2 describes the possible measured values that may result from a true concentration of zero. Due to random variability, half will be positive and half negative. Of most interest is how far away from zero in the positive direction those measurements are likely to fall. This can be described using the standard deviation of the data, relating the distance from the center to a statement of probability through the t-distribution. Measurements have about an 18 percent probability of falling at least one standard deviation above zero when the true concentration is zero, given that the standard deviation was computed using a sample size of 7 replicates. The probability of falling at least two standard deviations above zero, when the true concentration is zero, is about 5 percent. At a distance of three standard deviations, the probability is just over 1 percent. The choice of a distance to represent the detection limit is made so that no more than a small percentage of the measured values truly originating from a zero concentration will fall above the limit. As one example of setting a detection limit, USEPA (1982) describes the procedure for computing what it calls the method detection limit (MDL). Seven or more replicate analyses are performed on a standard where the chemical is present at a low concentration. The standard deviation (abbreviated s) of these measurements is then multiplied by 3.14, the one-sided t-statistic for a sample size of n=7 and probability of exceedance of 1 percent ($\alpha = 0.01$). This false exceedance rate is called the false positive or Type I error, the rate of measuring a value further to the right than the dashed line pictured in Figure 3.2. For a standard deviation s equal to 0.25, the detection limit (here USEPA's MDL) is set at a value of 3.14•0.25, or 0.78. Any measurement falling above 0.78 would be declared to have a concentration that is significantly different from zero at a 1% false positive rate, and the analyte is declared to be present in the sample.

There are several real-world complications that laboratories must deal with that make this process too simplistic. But the end result is that measured concentrations below the detection limit are not reported as individual values. The analyst does not

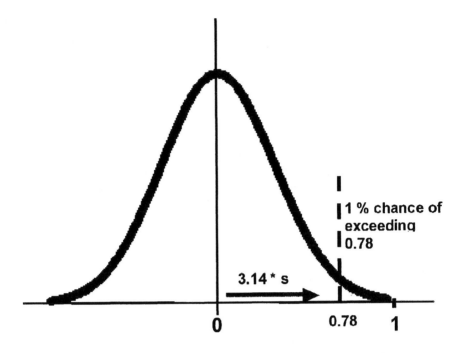

**Figure 3.2.** Setting a detection limit at 3.14 times the standard deviation above zero. This prevents false positives.
*Lowers the risk of*                    *PREVENTS ⇒ CERTAINTY*

consider the chemical to have been proven to be present in the sample. Instead, a censored value of "<RL" (where RL is the reporting limit) is stored in a database (see Chapter 4) and presented to the user. A helpful concept is to view these data as an interval rather than as a single number, with concentrations somewhere between 0 and the reporting limit. Such "interval censored" data can be incorporated into many statistical routines.

This definition of a detection limit assumes that the underlying laboratory variability can be described by a single number, the standard deviation. In practice, this variability changes from day to day, analyst to analyst, instrument to instrument, and from sample to sample as a function of the sample matrix. This is particularly true in a production lab, where any given sample entering the door may be assigned to one of a number of people and instruments. One approach to account for these changes in the quality of its measurements was devised by the U.S. Geological Survey and called the "Long-Term Method Detection Level" or LT-MDL (Oblinger-Childress et al., 1999). The standard deviation used to compute the LT-MDL incorporates the variability among the multiple instruments and multiple operators any given sample may be assigned to upon entering the laboratory. The LT-MDL is reevaluated each year, as equipment and operating conditions change. This and similar processes that more correctly track the precision of production laboratories make it even more likely that reporting limits will change over time (multiply-censored data). This in turn means that end-users must understand and use interpretation methods that correctly incorporate data having multiple reporting limits.

The process of censoring values below the detection limit and reporting them as a less-than value controls one type of error, the false positive rate. But there is a second type of error, a false negative, that is not addressed by this procedure. False negatives, also called Type II errors, are one motivation to use a higher reporting limit, the quantitation limit.

## The quantitation limit

Thresholds above which single numerical values (rather than an interval) are reported.

Quantitation limits arise from two distinct needs. The first need has been for a threshold above which reliable single numbers can be reported. These thresholds have been given names such as the "limit of quantitation (LOQ)" (Keith, 1992), and the "practical quantitation limit (PQL)". They are generally computed as about 10 times the standard deviation of a low standard such as the one used to define the method detection limit. The factor of 10 has been around for a number of years (USEPA, 2003), and a concentration 10 times the background variability is considered large enough by most chemists that a single number might be comfortably reported. The result is a threshold that is a little over 3 times the value of the detection limit $\left( \dfrac{10}{3.14} = 3.18 \right)$.

Quantitation limits are used because an analyst is uncomfortable reporting a single value for low measurements, given that the standard deviation is still somewhat large in comparison to the signal itself. For measurements between the detection and quantitation limits the analyst believes that there is a signal (the analyte is present at a trace amount), but that the signal is small in comparison to the variability of the measurement process. To avoid reporting an unreliable single number, the analyst constructs another threshold above which a single number may be reliably reported. Measurements above this threshold may be quantified by a single value; those below are usually not.

The second need identified by laboratory analysts is for a threshold that "protects against false negatives". A false negative occurs when a measurement whose true concentration is at or above the detection limit is reported as <DL. A sample whose true concentration is exactly at the detection threshold (0.78 in Figure 3.2) has a 50 percent chance of being recorded as <DL. This must be understood in the light that any measurement with a true concentration of X has a 50% chance of being reported as less than X, and a 50% chance of being reported as greater than X, assuming there is no bias (100% recovery). True concentrations of samples are never known, so there is a 50% chance of a "false negative" for every chemical measurement. But for the special case of the detection limit, some analysts consider the possibility of

a false negative to be a problem. A higher limit than the detection limit (as previously defined) is one method that has been used to address false negatives. Alternate terms for a threshold addressing this need have included the "method reporting level" (MRL) as well as the "detection limit" of Currie (1968) and the "reliable detection limit" of Keith (1992).

To determine a threshold that protects against false negatives (Type II errors), a probability distribution is set around the threshold, sliding the threshold value upwards until there is only a small likelihood that a measured value will fall below the detection limit (Figure 3.3). The probability distribution is assumed to be the same as that used for determining the detection limit. The result is a threshold 6 to 10 times the standard deviation. The probability of falsely reporting an analyte to be below the detection limit when it is truly at the higher threshold will be small. Assuming that the standard deviation of measurements at the higher limit is the same as the value computed using replicates at a low standard, the probability of a false negative is one percent when the higher limit is double the detection limit, or 6.28•s (Figure 3.3). So thresholds resulting from protection against false negatives are generally twice the detection limit. Some chemists will report single numbers above this limit, making this higher limit the effective quantitation limit. Most instead use the previously discussed 3.18 times the detection limit as the quantitation limit, and because it is higher than 2 times the detection limit it provides even greater protection against false negatives.

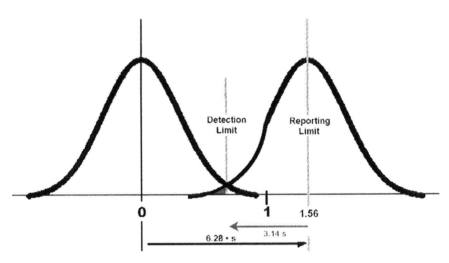

**Figure 3.3.** Establishing a higher limit at twice the detection limit, or 6.28 times the standard deviation, to protect against false negatives.

## False negatives

The concept of a false negative provides an important example of the difference in perspective between laboratories and data users. From a laboratory's point of view, a serious error is made if an individual sample whose true value is at or above the detection limit is reported as below the detection limit. The argument goes that a true concentration exactly at the detection limit has a 50% chance of being record-ed as below the detection limit, and so erroneously reported as a nondetect (Figure 3.4). To avoid this a higher threshold is instituted.

However, there are two counterarguments that lead to making no adjustment for false negatives. First, creating a higher threshold value does not in itself reduce the likelihood of false negatives. A true concentration at the higher quantitation limit has a 50% chance of being erroneously reported as a censored value below that limit. The limit is higher, but the effect is the same. In fact, for any concentration there is a 50% chance that the measured value will be below the true level, assum-ing 100% recovery. Creating a higher threshold does not in itself solve this prob-lem.

Second, if the true concentration were just a hairs-breadth below the detection limit, there is an almost 50% chance that the measured value will exceed the detec-tion limit, and so be erroneously reported as a detected value. The same error char-acteristics surround each true concentration. Looking at measurement errors from a collective standpoint, as a data user would do, the positive and negative errors will tend to balance out (Figure 3.5). Without adjusting for false negatives, the propor-

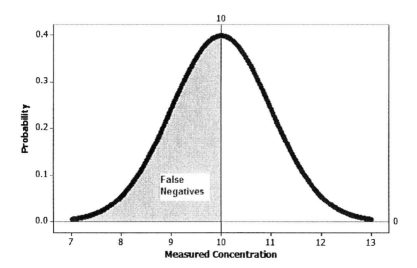

**Figure 3.4.** Probability of a "false negative" when the true concentration is at the detection limit of 10. Shaded probability equals 50%.

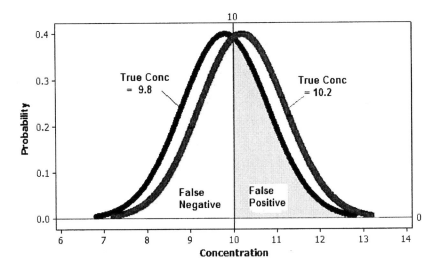

**Figure 3.5.** Balanced probabilities of false negatives and positives for true concentrations just below and above a detection limit of 10.

tion of values falling within each interval of concentration remains correct if each measurement has the same variability. The same percent of values near each boundary will by chance fall into a higher category as those falling into a lower category. So from a perspective of the overall data set, there is little need to censor data based on an avoidance of false negatives. They are balanced by errors in the opposite direction.

### Data between the limits

There is general agreement among laboratories that no numeric values should be placed on measurements below the lower threshold (the detection limit) because of the risk of false positives. Only a few scientists have advocated reporting these readings along with a measure of their analytical error, that is, $0.3 \pm 0.78$ (Porter et al., 1988). There is agreement that single values above the higher threshold (the quantitation limit) should be reported as measured. The controversy among practices of various laboratories is in what to do with measurements in the region between the two thresholds. Here the chemist generally believes that the analyte is present in the sample, but at concentrations that cannot be quantified with precision. Older analyses in this region were reported as simply "trace" or "detected". Not making the threshold value itself available is a serious problem. The large amount of information present in these data is contained in the proportion of values below specific thresholds, so reporting the numeric value of the threshold is crucial for capturing and analysis of this information. Sometimes the threshold value for an observation reported only as "trace" can be recovered by looking at lab records, or by using the lowest reported value in the database for that time. Recent analyses (say since 1980) generally do report a threshold value, usually the higher quantitation limit, resulting in a value of <QL for these in-between measurements. Most labs following USEPA guidelines will not report single numbers below the quantitation limit.

Controversy arises when data users advocate reporting numerical values between the detection and quantitation limits (Gilliom et al., 1984). Several labs have begun reporting these values, qualified by a remark that means something like "user beware" (ASTM, Sec. D4210, 1983; Oblinger-Childress, 1999). This remark indicates that the relative error for these measurements is high, and the individual values suspect. Most data users incorporate these values as if they were equivalent to values above the quantitation limit, effectively re-setting the detection limit as the reporting limit. Most laboratories would consider this risky, as calibration standards are usually not available or accurate in this range.

If data between the limits are not handled correctly a bias may be introduced. This bias is discussed in the next section on "informative censoring". Three methods for handling data between the limits without introducing a bias are:

1. Use the quantitation limit as the reporting limit. All values below the quantitation limit are considered nondetects labeled "<QL". Values between the limits are considered too unreliable to report as single numbers, and are reported as <QL, as are all values measured below the detection limit. Data analysis methods of this book may then be applied directly. However, measurements that signal presence of the analyte are lumped with measurements not distinguishable from zero. Some information is lost in comparison to the next two methods.

2. Use the detection limit as the reporting limit. This is the de facto result when data users take values reported between the thresholds as similar in precision to those above the quantitation limit, ignoring the qualifier. Values "<DL" are censored. Values between the limits are used as individual values. The advantage over method 1 is that measurements different than zero are recognized as higher than true nondetects. The risk is that there may be too much variability to reliably treat the data between the limits as anything other than tied with one another.

3. Use interval-censoring methods. Data between the limits are assigned tied ranks (nonparametric methods) or assigned to an interval (parametric methods) higher in value than those assigned to data below the detection limit. The ordering of data is preserved – the <DL group is considered lower than the in-between group – without assigning single values to observations in either group.

For an example of option 3, consider a data set with 7 values below the detection limit, 9 values measured between the limits, and 12 values measured and reported as individual numbers above the quantitation limit. There are 28 observations in all. The nonparametric approach to interval censoring assigns the average rank of all values below the detection limit to those measurements. All 7 values below the detection limit are assigned the average of ranks 1 through 7, or 4. So the 7 lowest observations have a rank of 4. Values between the two thresholds are similarly assigned the average of their ranks, recognizing that they are higher than values below the detection limit. The 9 values measured between the thresholds are assigned the average of ranks 8 through 16, or a rank of 12. Using the average of the possible ranks the data could have obtained preserves the sum of the ranks, which is a statistic used in nonparametric data analysis. Values above the quantitation limit remain the same as those they would have received if there had been no censoring, starting with the rank of 17 and going up to a rank of 28. More details on methods using nonparametric and parametric interval censoring are given in later chapters. Interval censoring in a parametric framework is described in Chapter 6, where an example of computing the mean is given.

The decision of which of these three methods to use is a decision that should be made by the data user in consultation with the laboratory scientist. Understanding of the relative precision of the data between the limits is key for determining how best to represent them. The three methods differ in the precision with which the lower end of the data distribution is represented. However, all three methods are unbiased in that the probability distribution (the percentiles) of data are not consistently shifted below or above their true values. An unfortunate result of the struggle to accurately represent censored data is that a fourth method is now being used to report censored data which does introduce a bias. This is a method that in other disciplines is called "informative censoring".

## Informative censoring – biasing interpretations

One of the assumptions behind all methods for interpreting censored data is that the measurement value does not influence the type of censoring process used. One does not decide to use one method of censoring for one range of concentrations, and another method for another range of concentrations. The process of censoring should be "non-informative" in regards to the concentrations themselves. When this assumption is met, the proportion of data below any threshold can be validly computed and compared to the proportion below another threshold. These proportions are another way of stating the percentiles of the data set, and represent the primary information present in data with censored values. Percentiles can be correctly computed and interpreted when the censoring mechanism is non-informative.

Informative censoring invalidates the computation of percentiles (and other summary statistics or tests). As an example, informative censoring arises in the medical sciences when the variable measured, the survival length of patients following diagnosis of a disease, influences how they are censored (Collett, 2003). Suppose the survival lengths of two groups of patients is to be compared, one group that has received a particular medical treatment, and another which has not. The goal is to determine if receiving the treatment is beneficial and generally lengthens a patient's lifespan. Censored values in these studies are patients who survive beyond the end of the study, and whose life spans are "greater thans". An example of informative censoring is the determination by a patient or doctor that an individual's expected survival time is short, and so they refuse (or are not considered for) treatment. In a feedback loop, persons who are projected to live longer are treated, and so the outcome that persons live longer with treatment cannot be ascribed just to the treatment. It may be due to the selection process itself.

A process similar to the feedback loop above is currently implemented by some laboratories in an effort to provide information their data users are requesting, while preserving a sense of protection from false negative (Type II) errors. One such process is the Laboratory Reporting Level (LRL) of the U.S. Geological Survey (Oblinger-Childress, 1999). Such processes have not yet been widely recognized as a problem, but as informative censoring they produce biased results as interpretation methods are applied to these data. Numerical values between the detection and quantitation limits are reported, though qualified. So for values measured between the limits, the detection limit is the effective reporting threshold. For observations measured as less than the detection limit, however, values are reported as less than the quantitation limit, <QL. These data are reported using the (higher) quantitation limit as the reporting threshold in order to avoid false negatives. The choice of reporting limit is therefore a function of the measured concentration of the sample. The result is that all interpretations of data reported in this manner will be biased. This is pictured in Figure 3.6.

Figure 3.6a shows the original measured concentrations as a bar graph. Suppose that forty percent of observations are measured between 0 and the detection limit (the white bar in Figure 3.6a). Twenty-five percent are measured between the detec-

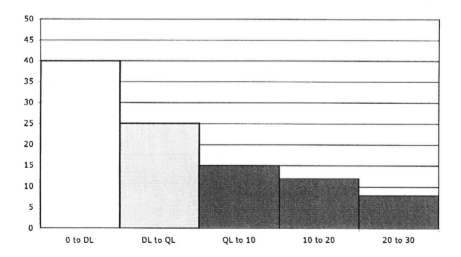

**Figure 3.6a.** Proportions of data within ranges of concentrations as originally measured.

tion and quantitation limits (light gray bar), and the remaining higher measurements reported as detected values (dark bars). The measurements between the limits are reported along with a qualifier that these observations are 'estimated', but still reside between the two limits. This bar graph applies as long as the values measured below the detection limit are reported as "<DL". Figure 3.6b shows the same data with informative censoring. The only difference is that values measured below the detection limit are now reported as being below the quantitation limit or "<QL", as if they might belong anywhere from zero up to the quantitation limit. The probability (forty percent) that observations may fall below the detection limit is spread evenly along the entire range from zero to the quantitation limit. This is pictured in Figure 3.6b as white bars totaling 40 percent evenly split between two categories, 20 percent of observations in each category. The result of informative censoring is that the probability that an observation might fall between the detection and quantitation limits is exaggerated, and the probability that it would fall below the detection limit is underestimated, in comparison to the proportions actually measured. The shape of the histogram has been changed, and so too will all interpretations that follow. This upward bias is picked up by any subsequent procedure, from the simplest computation of means or percentiles to more complex methods such as maximum likelihood.

Laboratories that attempt to satisfy data users while avoiding false negatives may fall into the trap of informative censoring. While arising from data users' requests for "numbers", every interpretation method they hope to employ is undermined. Instead of using an informative censoring mechanism, any of the three valid options of the previous section would avoid bias, and so avoid confusion as to the cause of

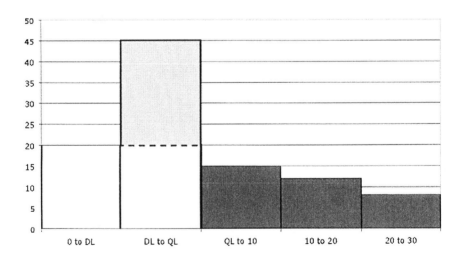

**Figure 3.6b.** Proportions of the 3.6a data within the same ranges after informative censoring. The lower end of the distribution has been shifted dramatically upward.

observed similarities or differences in the patterns of data. For labs that report data in this way, it is imperative that data users re-censor their data using one of the three unbiased methods listed in the previous section:

1. Censor all data below the quantitation limit as "<QL",
2. Censor only those data below the detection limit using the value "<DL", or
3. Use interval-censoring methods for interpretation.

### Limits when the standard deviation changes with concentration

Many papers in analytical chemistry have over the years modeled the variation of concentration as a function of concentration itself. Variation takes on a form something like plus or minus a percentage of the measurement, rather than plus or minus a constant number across the range of concentrations. Higher concentrations have higher variability. Logarithm or square root transformations of concentration are often used to convert measured concentrations to data approaching a more constant variance. Figure 3.7 is an example of data where the variability of measurements of prepared standards increases as the concentration of those standards increase. There are five measurements for each standard solution. The five replicates for a standard concentration of 3 overlap to the extent that not all five can be distinguished on the plot, while the five at higher concentrations show much more scatter.

Gibbons and Coleman (2001) state that this situation is common – variation increases as concentration increases. If this is true, detection and quantitation lim-

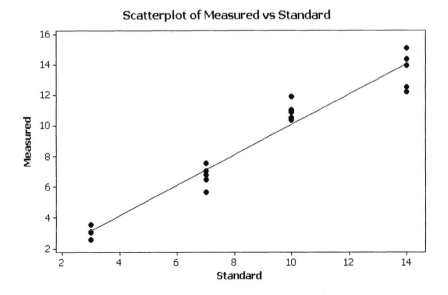

**Figure 3.7.** Variance of measurements increases with concentration.

its calculated assuming a constant standard deviation will often be inaccurate. Those computed using a higher concentration standard will obtain a higher estimate for the standard deviation. Projecting this higher standard deviation all the way down to zero, higher detection and quantitation limits are estimated than would be obtained with a lower concentration standard. Two laboratories that have the same underlying method precision will get different estimates for detection and quantitation limits when they use different concentration standards to set those limits.

This last statement should be of great interest to data users. Labs with the same precision characteristics should have the same detection and quantitation limits, but may set different limits due to the use of different standard concentrations. Censored observations from identical samples sent to these labs will be assigned different numbers if substitution is used! The variety of reporting limits resulting from causes unrelated to characteristics of a submitted sample is a strong reason to avoid substitution methods. The substituted number may be more strongly influenced by the concentration standard used two months prior to set the detection limit than it is by the concentration of the analyte present in the sample.

Gibbons and Coleman (2001) propose five steps for calculating censoring limits when the standard deviation varies with concentration (calling these "calibration-based limits"):

1. Measure the concentrations of replicates for several concentration standards ranging from very low to above the level expected for the quantitation limit.
2. Perform a weighted least-squares regression of measured concentration (Y) versus concentration of the standards (X). Weights may be computed as the inverse of the observed variance of the replicates at each concentration standard. Or they may be modeled using a regression of the variance versus the concentration standard values.
3. Guess a concentration for the detection and quantitation limits, and compute initial estimates of the standard deviation for those concentrations.
4. Use these initial estimates to compute initial values for the detection and quantitation limits.
5. Iterate between steps 3 and 4, alternatively computing estimates of standard deviation and subsequent censoring levels, until the estimates converge.

Using weighted least-squares, measurements with higher variability further from a concentration of zero have less influence on the final estimate of standard deviation than do those with lower variability closer to zero concentration. So if variability indeed increases with concentration, weighting produces a better estimate for the standard deviation near zero concentration, and therefore a better estimate of the detection limit, than does an assumption that the measurement variation is constant.

## For further study

A summary of the development and history of detection and quantitation limits from the perspective of one "school" is given by the USEPA (2003). Ideas to date on how to set analytical censoring limits are reviewed, though limits computed with a single standard deviation are emphasized. They also discuss both analytical and statistical issues raised by a variety of people involved with Clean Water Act determinations. This is primarily a review of how definitions of censoring thresholds have changed over time.

Currie (1968) first called for standardization of methods for computing censoring limits. His interest was in radiochemical data, but the concepts have since been adapted for use in aqueous environmental chemistry. Many practices today in environmental laboratories draw directly from the ideas presented in his article.

Gibbons and Coleman (2001) and Gibbons et al. (1997) present a detailed discussion of the process for setting detection and quantitation limits when the standard deviation is judged to be a function of the concentration value, rather than assuming that the standard deviation is constant between zero and the concentration standard. These concepts have been adopted by the American Society of Testing Materials (ASTM, 1991; ASTM 2000) and given the acronyms "interlaboratory detection estimate" or IDE and "interlaboratory quantitation estimate" or IQE. Several comments on the 1997 article have also appeared; see Kahn et al. (1998) and Rigo (1999) as well as the responses from the original authors that follow those comments.

Another paper by Gibbons (1995) on the deficiencies of current practices of setting detection and quantitation limits is followed by several discussion papers debating the points which Gibbons makes. Of special note is the discussion by White and Kahn (1995) who defend current USEPA practices. The original and discussion papers together provide a review of the current debate over the process of setting reporting limits.

# 4

# STORING DATA IN DATABASES

How might data that include nondetect values be stored in and retrieved from a database? Can they be stored in such a way as to make analysis easier? Goals for database storage of censored datasets might include:

- Censored and detected observations are both clearly identified and distinguished.
- Left-censored (nondetect) data can be distinguished from right-censored (greater-than) values.
- Data are easily incorporated and used by statistical software.
- Censoring by detection versus quantitation limits can be distinguished and recorded.

On paper, designating a value as a nondetect is trivial – a text character is used, the less-than sign (<). It would seem that designating nondetects within a digital data set should be no more difficult. However if the character "<" is placed into a database or spreadsheet cell, the entire entry is considered text and therefore not available for computations. Also, statistical software cannot use the less-than character to perform calculations. Another method of data storage is required.

There are three commonly-used methods for storing data with nondetects:

1. Negative numbers
2. Indicator variable
3. Interval endpoints

Each is discussed in turn. The first method is simple but dangerous. It meets only the first goal of identifying detected and nondetected data. The indicator variable and interval endpoints methods go much further toward meeting all the storage goals listed above. An indicator variable is most often used by nonparametric procedures in statistical software, particularly the Kaplan-Meier and related methods. Interval endpoints are most often used by parametric methods for maximum likelihood and, in fact, are crucial for designating 0 as a lower bound for these methods. The data analyst will often have to use both the indicator variable and interval endpoint methods for the same data set in order to compute both parametric and nonparametric analyses.

## Method 1. Negative numbers

To circumvent the problem of using the < sign in a numeric database some data users substitute the minus sign (–) , so that all nondetects appear as negative numbers. A <1 is represented as a –1, and so on. This method is an efficient use of space in a database, requiring no more than would be needed without censoring. However, it has far more disadvantages than advantages. The method can only be used when the original data are completely non-negative (most but not all environmental data are so constrained). The most serious deficiency is that negative values could be mistakenly interpreted as true values below zero by persons unaware that this method was used. In that case summary statistics and other analyses performed directly on these data with no adjustment would be strongly biased. For example, a simple mean of data that include values which should be considered between 0 and 10 (<10), but is computed using values of −10, will be far too low.

Therefore it is best to avoid this format for storing censored data.

### *Example*

Given three observations, concentrations of <1, <5 and 10, a spreadsheet would record one column having three entries:

Concentration
− 1
−5
10

## Method 2.  Indicator variable

Using an indicator variable, two values are required to represent each observation. The first variable stores the measured values for detected observations, as well as values for the detection limits of all nondetects. The second variable contains a code that indicates whether the value in the first column is a detected concentration or the detection limit for a censored observation. Typically a 0 is used to indicate one state, and a 1 for the other. In this book a value of 0 will generally designate a detected observation, while a 1 designates a nondetect. Some packages also allow text phrases such as "censored" and "uncensored" or "above" and "below" to be stored in the indicator variable. Text phrases remind the data user which indicator values correspond to which type of observation. If text phrases are not allowed, the variable name for the 0/1 variable should remind the user which state refers to which number. A name such as "BDL1" indicates that below detection limit data have values equal to 1. Data with BDLI values equal to zero must then be detected values. If the remark variable is ignored, summary statistics will be biased high, because the detection limit will be erroneously considered as the measured value for all nondetects. This bias is less severe than mistakenly using negative numbers as observed values rather than as an indicator of a nondetect, but it remains a serious error.

Indicator variables have been used to store multiple text phrases, often single letters, to designate different types of remarks about the data. One letter may indicate that values are below a detection limit, while another used for values between the detection and quantitation limits. A different letter could indicate a "greater than" value. Letters must be translated into ranges of numerical values within which each observation falls, such as 0 to 1 (below a detection limit of 1 µg/L) and 1 to 3 (between the detection and quantitation limits) in order to use the data in a statistical routine.

The indicator variable method is able to meet all four of the listed goals for storage of nondetects if multiple indicators are used to distinguish data between the limits from true nondetects, and greater-thans from less-thans. The simpler binary 0/1 indicator is only able to meet the first and third goals for data storage. The method's primary disadvantage is that it can sometimes be difficult to remember which state each indicator value refers to.

### Example, cont.

The three values of <1, <5 and 10 are represented by two columns below. A 1 indicates a censored observation.

| Col 1. Concentration | Col 2. Indicator |
| --- | --- |
| 1 | 1 |
| 5 | 1 |
| 10 | 0 |

## Method 3.  Interval endpoints

If databases cannot store or assign text to numerical values, this third method is the easiest and the least confusing way to store censored data.  It is also a format that represents most closely how software uses censored data.  All values in the data set are represented by the interval those data fall within.  Endpoints for the interval, high and low, are stored in two variables.  For detected observations the values of the two variables are identical.  For censored data the values differ.

Interval endpoint storage is a flexible storage system that meets all four of the goals for data storage listed previously.  Left-censored data (less-thans) can be distinguished from right-censored (greater-thans) data, so that both may be possible within the same data set.  The comparison of values in the first variable to those in the second variable determines whether the observation is above the detection limit, or a nondetect, or a greater-than. The format is illustrated below.

### Example, cont.

For the data set consisting of a <1, <5 and a 10, two variables represent each observation, one at the lower end (Start) of an interval within which the measured values lie, and a second at the upper end (End) of the interval.  For left-censored data a value of 0 is entered in the Start variable to denote that the nondetect is no lower than zero, and the detection limit is entered as the largest possible value in the End variable.  This method clearly shows that the range of values does not go negative. The detected value of 10 has the same number for both variables.

| Start | End |
| --- | --- |
| 0 | 1 |
| 0 | 5 |
| 10 | 10 |

If it is possible for data to extend in value out to infinity or minus infinity, an asterisk (*) is placed in the appropriate column to represent infinity. The asterisk states that there is no known boundary, unlike the 0 lower boundary in the Start variable above. A right-censored value of >100 would be designated as:

| Start | End |
|-------|-----|
| 100   | *   |

where the asterisk denotes that no specific upper limit is known for "greater-than". Where there is a boundary, such as with a lower bound of 0, that boundary should be used. Maximum likelihood methods (discussed later) will produce different results depending on whether 0 or an * is found in the Start variable. Entering an asterisk for the Start variable produces estimates that are too low because values below zero are considered possible by the procedure when they actually are not possible for bounded environmental data.

Observations with true negative values can be represented using the interval endpoints format. The format is entirely numeric – no text is needed to designate censoring. Reading a direct printout of the data set is clear – it is perhaps the least confusing of the three methods. However if the data are not recognized as being censored, summary statistics of either variable alone will give erroneous answers; the Start variable will be biased low and the End variable biased high.

# Exercises

All data sets for the exercises in this book are found on the website

**http://www.practicalstats.com/nada**

in both Minitab® (*.mtw) and Excel (*.xls) file formats. Also found there are all Minitab® macros (*.mac) used throughout the book for computing the in-text examples and exercises.

**4-1**    Millard and Deverel (1988) measured copper and zinc concentrations in shallow ground-waters from two geological zones underneath the San Joaquin Valley of California. One zone was named the Alluvial Fan, the other the Basin Trough. Their data are found in the data set CuZn (use CuZn.xls if using software other than Minitab®). In addition to the two columns of concentrations, there are paired columns in the Indicator Variable format designating which of the observations represent detected concentrations, and which are "less-thans". The indicator variable names (CuLT=1 and ZnLT=1) show that "less-than" observations have a value of 1, while detected observations are indicated by a 0.

Create two new variables in the Interval Endpoints format, StartCu and EndCu, that will contain the same information given by the current variables Cu and CuLT=1.

**4-2**    What problem may have occurred with the following censored data set? What characteristics lead to that conclusion?

0.55 0.6  0.8 0.85 0.9 <1  <1  <1  <1  <1 1.0 1.2 1.7 1.8 2.2 2.6 3.5

**4-3**    Flip the copper concentrations for the Alluvial Fan zone to a right-censored format and store in a new variable named something like "FlipCu". Plot both Cu and FlipCu with either a boxplot or histogram (ignoring the less-than indicators at this point). How do the plots of the two variables compare? Given that the variable Cu is skewed, take logarithms and repeat the process.

# 5

# PLOTTING DATA WITH NONDETECTS

Methods for plotting uncensored data can be extended to censored data with only a few modifications. Boxplots can illustrate the distribution (shape, typical values, outliers) of censored data. Probability plots can provide a visual check of conformance to a specific distribution such as the normal. An empirical distribution function (edf) depicts the distribution function of data with more precision, but perhaps less intuition, than does a boxplot. A survival function plot is an edf specifically developed for censored data. Each of these types of plots will be illustrated using the June atrazine concentrations of the Atra data set, which includes values below one detection limit.

## Boxplots

Boxplots (Helsel and Hirsch, 2002, Chapter 2) are one of the most intuitive ways to visualize a data set. They employ three percentiles ($25^{th}$, $50^{th}$ and $75^{th}$) that define the central box. The relative positions of the percentiles show the center, spread, and skewness of the data. Boxplots also represent outliers as individual points. Nondetects can be incorporated into boxplots by using the information in the proportion of data below the highest detection limit. Boxplots should never be drawn by deleting nondetects, drawing the graphic using only detected values. Deleting nondetects destroys all meaning of the percentiles of the data set, which is what the box of a boxplot represents. There is a better way.

To draw a boxplot for data with a single detection limit, all nondetects are set to any single value lower than the limit. A horizontal line is drawn at the detection limit. All detected values will be represented correctly, but the distribution below the detection limit is unknown and should not be represented in the same way as the portions above the limit. The portions of the box below the detection limit line are often removed. This "boxplot at sunrise" (Figure 5.1) is an accurate representation of the information contained in the data set. The proportion of data censored is indicated by how much of the graphic is below the horizon. In Figure 5.1 the lack of a lower line for the box, but presence of the central median line, shows that between 25 to 50 percent of the data are nondetects (there are 9 of 24 values or 38% below 0.01 for the atrazine data set).

For multiple detection limits, only data above the maximum detection limit is known exactly. Portions of the box above this limit are drawn with solid lines. In the most conservative approach, everything below the maximum detection limit would not be shown (Figure 5.2). However, if too much of the box is invisible below the horizon, the 25th, 50th and 75th percentiles can be estimated if necessary and drawn with dashed lines (Figure 5.3). Percentiles are estimated using either

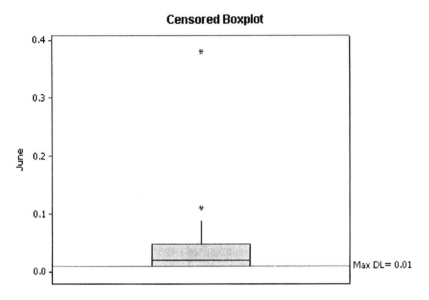

**Figure 5.1.** Boxplot of censored atrazine data. The proportion of censored data is between 25 and 50 percent, as shown by the presence of a line for the 50th, but not the 25th, percentile.

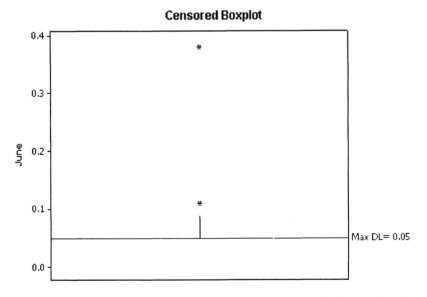

**Figure 5.2.** Boxplot for altered atrazine data with detection limits at 0.01 and 0.05. The 25th, 50th and 75th percentiles are all below the maximum detection limit and are not shown.

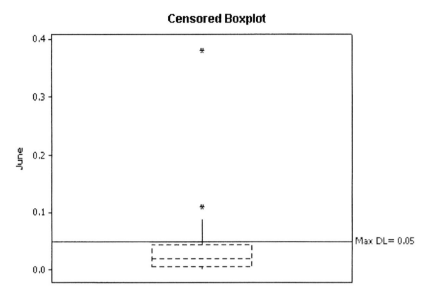

**Figure 5.3.** Boxplot for altered atrazine data with detection limits at 0.01 and 0.05. The 25th, 50th and 75th percentiles have been estimated using the robust ROS method of Chapter 6.

Kaplan-Meier or robust ROS (see Chapter 6). These methods incorporate the proportion of observations occurring below each detection limit when calculating percentiles.

**Histograms**

Histograms are not particularly useful plots for depicting censored data. This is because there is not one histogram that is unique to a data set – many equally valid histograms might be drawn from the same data. In Figure 5.4 the censored atrazine observations are drawn with their own (dark) bar to show that 38% of the data set is below the detection limit. For data with multiple detection limits the bar would include all data below the highest limit. Above that limit, detected observations are categorized into ranges and the percentage in each category shown (white bars in Figure 5.4). The detection limit (maximum detection limit for multiply-censored data) is represented by a vertical line in Figure 5.4.

**Probability plot**

Probability plots check the similarity of data to a normal or other specified distribution. The distribution is represented on the plot as a straight line, and observations

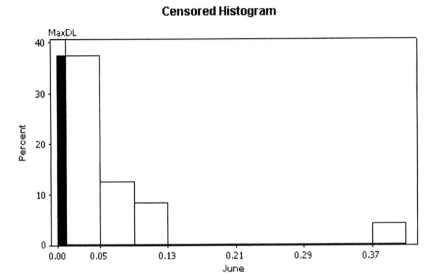

**Figure 5.4.** A histogram of the June atrazine data.

are plotted individually. If observations closely follow the straight line, they appear to closely follow the specified distributional shape. For censored data it is difficult to represent nondetects with a point because the location at which to place them is unclear. Substitution of numbers for nondetects would not work well as the shape of the plot will change depending on the values substituted. Worse yet, some casual readers will then miss the fact that no single numerical values are actually available for these data. However, the proportion of data below each detection limit can be computed in order to determine the placement of a line for each detection limit on the plot. Detected observations can then be plotted individually, including those that fall between detection limits. All detected values above the highest detection limit have probabilities (percentiles or "normal quantiles") that are known exactly. A probability plot for the June atrazine data is shown in Figure 5.5. Note that no individual points are plotted for nondetects. Based on values above the detection limit, the atrazine data appear to reasonably fit a lognormal distribution (the straight line).

For multiple detection limits, statistical software will assign percentile values to detected observations while taking into account the proportion of nondetects below each detection limit. All detected observations can be plotted, even those located between detection limits. Either Kaplan-Meier or robust ROS (see Chapter 6) is used to estimate percentiles for the detected observations. A probability plot for data with detection limits at 1 and 10 is shown in Figure 5.6. Note that three observations between 1 and 10, measured during the period when the detection limit was 1, are plotted.

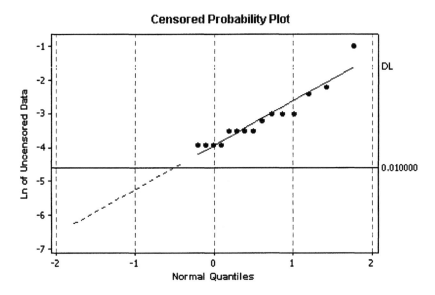

**Figure 5.5.** Probability plot for the censored June atrazine data.

## Empirical distribution function

A plot of the empirical distribution function (edf), also called a quantile plot, shows the sample percentiles (quantiles) of each observation in the data set. Edfs are sample approximations of the true cumulative distribution function (cdf) of a continuous random variable. The vertical axis lists the quantiles (median = 0.5 quantile, or 50th percentile) ranging between 0 and 1 (or 100%). The horizontal axis covers the range of numerical values of the data. Data points are plotted in sequence from low to high, and connected by straight lines to form the graph. By selecting one percentile value (say 50%) and reading across to the curve, the percent of observations below that value in the data set is obtained.

To construct an edf, the data are ranked from smallest to largest. The smallest value is assigned a rank $i = 1$, and the largest a rank $i = n$, where n is the sample size of the data set. Ranks are converted to percentiles using a "plotting position" p, where $p = i/n$. When data are tied, as for nondetects, each is assigned a separate plotting position (the plotting positions are not averaged). Tied values are seen as a vertical "cliff" on the plot, like the one in Figure 5.7 for atrazine data at the detection limit of 0.01.

The plotting position for an edf is an estimated percentile, the probability of being less than or equal to that observation. The largest observation has $i/n = 1$ and so is the 100th percentile, having a zero probability of being exceeded. For samples taken as a subset of the total population, it would be wise to recognize that there is a likelihood of exceeding the largest value observed to date. This can be represent-

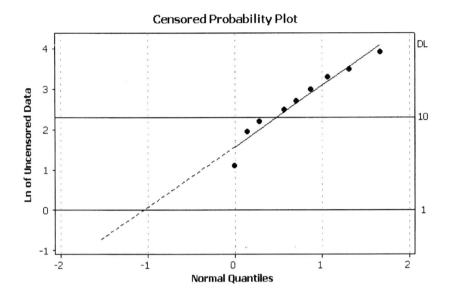

**Figure 5.6.** Probability plot for logarithms of censored data with two detection limits (at concentrations of 1 and 10).

ed on the graph by using a plotting position less than i/n on the vertical axis, though this is not done by standard statistical software.

Different plotting positions are used depending on the purpose and the tradition of a procedure. Numerous plotting position formulas have been suggested (Helsel and Hirsch, 2002), many having the general formula

$$p = (i - a) / (n + 1 - 2a)$$

where a varies from 0 to 0.5. Each differs in the probability of exceedance above the largest observation.

Five of the most common formulas are:

| Name | a | Formula |
|------|---|---------|
| Weibull | 0 | $i / (n + 1)$ |
| Blom | 0.375 | $(i - 0.375) / (n + 0.25)$ |
| Cunnane | 0.4 | $(i - 0.4) / (n + 0.2)$ |
| Gringorten | 0.44 | $(i - 0.44) / (n + 0.12)$ |
| Hazen | 0.5 | $(i - 0.5) / n$ |

The Blom plotting position is used most often on probability plots to compare data to a theoretical normal distribution.

Commercial statistical software is not available to draw edfs when there are multiple detection limits. Instead, percentiles are estimated using Kaplan-Meier meth-

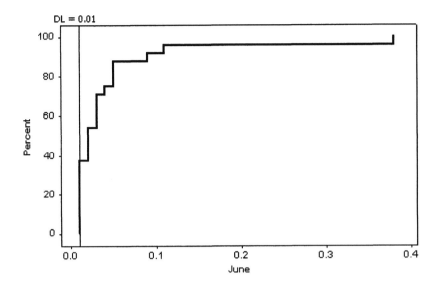

**Figure 5.7.** Empirical distribution function (edf) of the June atrazine data.

ods and plotted on a "survival function plot", which is similar to an edf.

**Survival function plots**

Survival function plots (Figure 5.8) are a modified edf. They are essentially an edf flipped side-to-side, plotting probabilities for the largest observations (in the original scale) to the left. In Figure 5.8 the detected atrazine observations are plotted as open circles. The atrazine scale was added to the bottom of Figure 5.8 for reference, and shows the largest values at the left side of the plot. Unless added with graphics software, survival plots will not show the original scale for environmental variables on the x-axis. Instead the original data have been flipped and labeled as "Time" or "Survival time". Built for disciplines that analyze right-censored (greater-than) data, survival plots will be harnessed to instead illustrate the edf of left-censored environmental data with multiple detection limits.

The survival function presents the probabilities of exceeding a value of "Survival time". Figure 5.8 shows a 97% probability of exceeding a "time" of 0.62. Time was created by subtracting atrazine from a value of 1.0, so that this point is also an atrazine concentration of (1−0.62) = 0.38 μg/L. The concentration of 0.38 represents the 97th percentile of the atrazine distribution, and so has a 97% probability that atrazine will be less than or equal to 0.38. Percentiles plotted on the y-axis are calculated with the Kaplan-Meier method described in Chapter 6.

Left-censored environmental data can be converted to right-censored "Time"

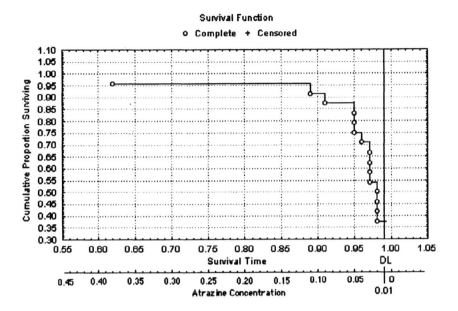

**Figure 5.8.** Survival function plot of the June atrazine data, with an additional Atrazine Concentration scale added for reference at the bottom.

data by subtracting each observation from a number larger than the maximum in the data set (see Chapter 2). This changes the "Atrazine Concentration" scale into the "Survival Time" scale of flipped data. In Figure 5.8 a survival time of 0.99 corresponds to an atrazine concentration of 0.01, the detection limit. The jagged survival-function probability line intersects the vertical detection limit at a "Cumulative proportion surviving" or probability of 0.38. Survival probabilities, equal to percentiles of the original observations, are found on the y-axis. There is a 38% probability of "surviving" to a value greater than 0.99, as well as a 38% probability of being less than the detection limit, just as previously calculated. Only the uncensored data are plotted on a survival function, but their positions are influenced by the censored as well as the uncensored observations. Further detail on Kaplan-Meier computations is given in Chapter 6.

## X-Y scatterplots

Scatterplots compare the values of two continuous variables, usually denoted X for the variable plotted along the horizontal axis and Y for the variable on the vertical axis. The paired X-Y values are visually inspected for patterns of correlation or trend – are values of Y predictably high or low for given values of X? A dilemma comes when either variable is censored – what numeric value should be used to place that observation on the plot?

Unfortunately, the most common practice is to substitute one-half the detection limit and plot the fabricated data as if it were measured there. The result is a false impression that these values are actually known, and that they are in all cases the same number. Neither is true. For multiple detection limits, plotting a fabricated value gives a false impression of the comparison between observations. A value of <10 plotted as a 5 is shown as if it were larger than a <3, when in fact the reverse might be true. All of the disadvantages of substitution in numeric procedures carry over into scatterplots. A signal that is present may be obscured. A signal may be shown which in reality does not exist.

Since censored values are known only to be within an interval between zero and the detection limit, representing these values by an interval, such as with a line segment, provides a visual picture of what is actually known about the data. Figure 5.9 shows a scatterplot of dissolved iron concentrations in summer samples from the Brazos River, TX, reported by Hughes and Millard (1988). They investigated whether trends occurred in iron concentrations over a 10-year period. Iron concentrations were censored at two detection limits during the study, at 10 μg/L n the earlier years and at 3 μg/L in later years. Detected observations are shown as single points, as usual. Nondetects are shown as dashed gray lines between zero and the detection limit.

The use of an interval conveys the uncertainty present in censored values. The lines are best shown as grayed-out rather than fully dark to emphasize that any one location within the interval is less likely than the single location shown for each detected observation. If the total mass of ink used for the interval could equal the mass used for a single detected point, the correct impression would be made. Figure 5.10 is another example, showing censored observations as an interval with jittered dots to represent uncertainty. In contrast, detected observations are shown as solid squares.

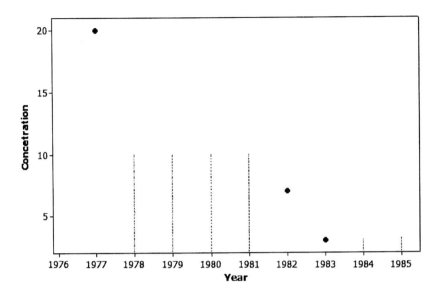

**Figure 5.9.** Scatterplot of dissolved iron concentrations over time. Nondetects shown as dashed lines. Data from Hughes and Millard (1988).

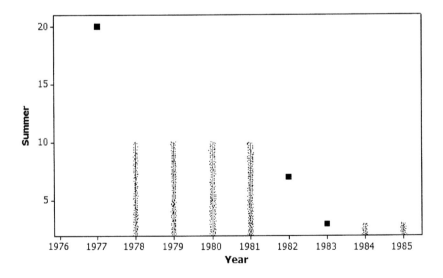

**Figure 5.10.** Alternate scatterplot of dissolved iron concentrations over time. Nondetects shown as jittered intervals. Data from Hughes and Millard (1988).

# Exercises

**5-1**     Plot a censored boxplot and censored histogram for Millard and Deverel's (1988) zinc concentration data found in the data set CuZn.  Use the Minitab® macros chist.mac and cbox.mac.  The zinc concentrations will need to be split into two columns, one for each zone, to plot the censored histograms.  The censored boxplots can be plotted with one command:

%cbox c3 c4 c5     (or abbreviating,  %cbox c3-c5)

and a box will be drawn for each group listed in column c5.

Describe the results – what characteristics of the data will likely be important for further data analysis?

**5-2**     The atrazine data used in this chapter are found in the data set Atra. Draw an empirical distribution function plot (edf, also called a cdf) for the June atrazine data.  In Minitab® this is done using the

Graph > Empirical cdf

command.  Using the Distribution dialog box, select "lognormal" as the best fitting distribution.  A lognormal distribution will be plotted as a blue line, and the empirical cdf with a red step function.  Compare the resulting plot to the survival function plot of Figure 5.8.  How are the two plots related?

# 6

# COMPUTING SUMMARY STATISTICS

More articles have been written comparing and recommending methods to compute summary statistics than for any other type of analysis of censored environmental data. As early as 1967, Miesch recommended use of the MLE for computing estimates of mean abundances (mass) of metals with censored measurements in rock samples (Miesch, 1967). Nehls and Akland (1973) recommended using one-half the detection limit to compute summary statistics for air quality data. Gilbert and Kinnison (1981) applied several methods, including probability plotting procedures, to censored radiochemical data. Helsel (1990) recommended the use of survival analysis methods for censored data, though the idea of transforming left-censored data to right-censored values had already been published by Ware and DeMets (1976). Use of each of these methods continues today. A review of several papers testing these methods and variations thereof for estimation of means, medians, variances, and other parameters is found at the end of this chapter. First, however, the primary methods available for computing summary statistics of censored data are introduced. Methods are classed into the three approaches of Chapter 2 – substitution, maximum likelihood, and nonparametric methods. A robust procedure based on probability plots is also included. Each method is illustrated using the same dataset, arsenic concentrations in urban streams on the island of Oahu (Tomlinson, 2003). Following the detailed description of these four methods, articles comparing their performance under a variety of conditions are reviewed and summarized.

Consider 24 arsenic concentrations (in μg/L) from urban streamwaters found in the *Oahu* data set.

| 0.5 | 0.5 | 0.5 | 0.6 | 0.7 | 0.7 | <0.9 | 0.9 | <1.0 | <1.0 | <1.0 | <1.0 |
| 1.5 | 1.7 | <2.0 | <2.0 | <2.0 | <2.0 | <2.0 | <2.0 | <2.0 | <2.0 | 2.8 | 3.2 |

Three detection limits are listed, along with detected observations below the lowest detection limit. Perhaps a more precise method with detection limit below 0.5 was put into effect in the latter part of the study? If so, the lowest six samples were measured using this method, and no nondetects were recorded during this time. The value of that detection limit was not stored with these data. Summary statistics for these messy data are computed by each of the methods that are discussed in the following sections.

## Substitution (fabrication) methods

Substitution remains the most commonly used method for computing summary statistics of censored environmental data. This is unfortunate, based on its poor performance in essentially all of the comparison studies reviewed later. Estimates for summary statistics could be computed by substituting any value between zero and the detection limit for each nondetect. Table 6.1 reports summary statistics resulting from substituting zero, one-half the detection limit, and the detection limit for each nondetect in the Oahu data. Figure 6.1 shows boxplots for the three sets of data after substituting these values. Estimates of mean and percentiles are quite different for the three substitutions, and would differ still if any other proportion of the detection limit were substituted instead; for example, 0.75•DL has often been used in geochemistry (Sanford et al., 1993) and $DL/\sqrt{2}$ in air quality and industrial hygiene applications (Hornung and Reed, 1990). Regardless of the value chosen, it is extremely unlikely that all of the samples recorded as a nondetect actually did occur at the same value. Therefore estimates of variability (standard deviation and variance) following substitution are unlikely to be close to their true values had no censoring occurred. Any substitution between zero and the detection limit is arguably as valid as any other. Yet after substituting different values, results may differ so substantially that decisions to remediate, or legislate, or prosecute, may change depending on the value substituted. For the Oahu example, if legal compliance were judged by whether the mean arsenic concentration remained below 1 μg/L, substituting 0 for nondetects would result in a judgment of compliance, while substituting the detection limit would not (Table 6.1). Using values in between might or might not result in compliance. None of the substitutions could be argued to be more valid than another. In this case it is clear that substitution must be abandoned – it is not an unequivocal method of analysis. There are better ways.

**Table 6.1.** Substitution results for Oahu arsenic data.

| Value Substituted | Mean | StDev | Pct25 | Median | Pct75 |
|---|---|---|---|---|---|
| Zero | 0.567 | 0.895 | 0.000 | 0.000 | 0.700 |
| 1/2 dl | 1.002 | 0.699 | 0.500 | 0.950 | 1.000 |
| dl | 1.438 | 0.761 | 0.750 | 1.250 | 2.000 |

**Maximum likelihood estimation**

Maximum likelihood estimation requires the assumption that a distribution (normal, lognormal or some other distribution) will closely fit the shape of the observed data. A mean and standard deviation for the distribution are computed based on the observed detected values, and the observed proportions of data below one or more censoring thresholds. Optimization of the mean and standard deviation produces the specific distribution that best fits the observed data.

In the late 1950s and early 1960s, several papers by A. C. Cohen introduced MLE for determination of the mean and variance of censored data. The method was fairly computer-intensive, beyond the computing power available to most people at that time. So Cohen developed a version which uses a lookup table to estimate the mean

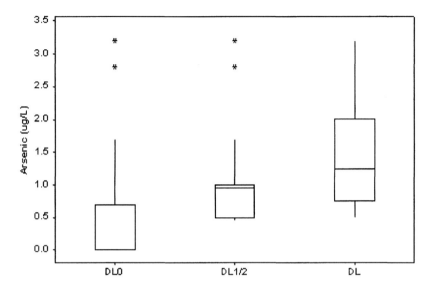

**Figure 6.1.** Boxplots of arsenic concentrations in an urban stream, Oahu, HI. Identical data except that nondetects are filled in with zeros (DL0), one-half the detection limit (DL1/2), and the detection limit (DL). Each is equally (un)likely.

and variance of a singly-censored (one detection limit) normal distribution by adjusting downwards the statistics for uncensored (detected) observations in response to the amount of censoring (Cohen, 1959). He presented an expanded lookup table with more detail in a subsequent article (Cohen, 1961). Though used by Miesch (1967) to estimate statistics of geochemical data, the method was not popularized for environmental sciences until Gilbert's 1987 book on environmental pollution monitoring.

Cohen's table-adjustment method is unnecessary today, since more accurate and versatile solutions of the likelihood equations are possible with modern statistical software. It has one serious drawback – the tables are restricted to the case of one detection limit. Most environmental data today contain multiple limits. These are easily handled by MLE methods available in commercial statistical software, but not by Cohen's method. However, the table-adjustment method is still sometimes recommended (USEPA, 2002a), and was considered 'new' in some fields as late as 1990 (Perkins et al., 1990).

Environmental data are more often similar to a lognormal than to a normal distribution, so the mean and variance of the logarithms are more typically estimated by MLE, whether with the table adjustment or by direct solution, and subsequently reconverted to estimates in original units. The traditional formulae for reconversion are derived from the mathematics of the lognormal distribution, and are found in many textbooks, including Gilbert (1987) and Aitchison and Brown (1957):

$$\hat{\mu} = \exp(\hat{\mu}_{\ln} + \hat{\sigma}^2{}_{\ln}/2) \tag{6.1}$$

$$\hat{\sigma}^2 = \hat{\mu}^2 \bullet \left[\exp(\hat{\sigma}^2{}_{\ln}) - 1\right] \tag{6.2}$$

$$C.V. = \left[\exp(\hat{\sigma}^2{}_{\ln}) - 1\right]^{1/2} \tag{6.3}$$

where $\hat{\mu}_{\ln}$ and $\hat{\sigma}^2{}_{\ln}$ are estimates of the mean and variance, respectively, of the natural logarithms of the data. These equations will work reasonably well if the data are close to lognormal in shape, and if the estimates in log units ($\hat{\mu}_{\ln}$ and $\hat{\sigma}^2{}_{\ln}$) are close to their true values. However, for small samples the estimates are typically poor enough to bias estimates in original units (Cohn, 1988), leading to overestimation of the mean and variance. MLE methods have not been found to work well for estimating the mean or variance of small (n < 30) samples in any of the papers reviewed later, particularly those assuming a lognormal distribution.

Estimates for percentiles are obtained by computing the percentiles in log units, assuming that the logarithms follow a normal distribution, and then retransforming. The kth percentile is therefore computed as:

$$p_k = \exp\left(\mu_{\ln} + z_k \sigma_{\ln}\right) \tag{6.4}$$

where $p_k$ is the kth percentile value in original units, and $z_k$ is the kth percentile of a standard normal distribution. For the median, k=0.5 and $z_k = 0$, so that $p_{0.5} = \exp\left(\mu_{\ln}\right)$. The exponentiated mean of the logarithms is sometimes given a special name, the geometric mean. When the logarithms of data follow a normal distribution, the geometric mean estimates the median of the data's original units (and not the mean).

MLE estimates can be thought of as the statistics of the distribution most likely to have produced the observed data, both censored and uncensored, given that the underlying process follows the assumed distribution. Checking this distribution assumption before computing estimates is a crucial first step. One method for checking the shape of a distribution is the probability plot (see Chapter 5). Figure 6.2 is a probability plot for the Oahu arsenic data. Uncensored values are plotted as solid circles, and a lognormal distribution with the same mean and standard deviation as the data is represented by the straight line. The data points follow a straight line pattern reasonably well for all but the lowest concentration. Therefore an assumption of a lognormal distribution for these data should produce reasonable estimates for summary statistics.

### Cohen's table adjustment method – example
To illustrate how Cohen's table adjustment method is computed, the procedure of Gilbert (1987) is applied to the Oahu data. In order to do this, all data below the highest detection limit must be considered censored, as the method works only for

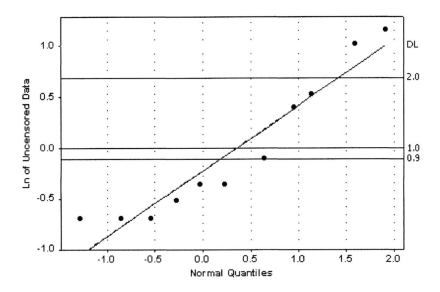

**Figure 6.2.** Lognormal probability plot for the Oahu arsenic data.

one detection limit. In other words, all values below 2 become <2, resulting in a data set of:

<2.0  <2.0  <2.0  <2.0  <2.0  <2.0  <2.0  <2.0  <2.0  <2.0  <2.0  <2.0
<2.0  <2.0  <2.0  <2.0  <2.0  <2.0  <2.0  <2.0  <2.0  <2.0   2.8   3.2

1. Compute h, the proportion of measurements censored. For the above data, h = 22/24, or 0.917. 91.7% of the observations are censored.
2. Compute the mean and variance from the uncensored observations. Given that most concentration data more closely follow a lognormal than a normal distribution, first convert the data to natural logarithms and then compute the $mean_u$ and $variance_u$ of the uncensored observations:

$$mean_u = 1.096 \quad \text{and} \quad var_u = 0.0089 \quad \text{(log units)}.$$

3. Compute $\gamma = \dfrac{var_u}{(mean_u - DL)^2}$ , where DL is the detection limit (in log units).

$$= \frac{0.0089}{(1.096 - 0.693)^2} = 0.0549$$

4. Estimate $\lambda$ from the table in either Cohen (1961) or Gilbert (1987). For h = 0.90 and $\gamma = 0.05$, $\hat{\lambda} = 3.314$.
5. Estimate the mean and variance of the log-transformed data.

$$\mu_{ln} = mean_u - \lambda(mean_u - DL) \quad = 1.096 - 3.314(0.403) \quad = -0.24$$

$$\sigma^2{}_{\ln} = \text{var}_u + \lambda(\text{mean}_u - \text{DL})^2 = 0.0089 + 3.314(0.162) = 0.547$$

6. Estimate the mean and variance in original units using equations 6.1 and 6.2:
$$\hat{\mu} = \exp(-0.24 + 0.547/2) = 1.034$$

$$\hat{\sigma}^2 = (1.034)^2 \bullet [\exp(0.547) - 1] = 0.778$$

For a given data set and assumed distribution, differences in summary statistics between Cohen's table-lookup estimates and MLE estimates using statistical software can be attributed both to the approximations built into the lookup table, and to the additional censoring required by table lookup for data sets with more than one detection limit.

### MLE using statistical software – example

MLE methods are available in commercial statistics packages. For environmental data it is important that interval endpoints can be specified, so that the lower bound of 0 can be represented. Minitab's®

### Stat > Reliability/Survival

menu selection allows a lower bound to be entered for left-censored data for its MLE procedure. The appropriate option is

### Parametric Distribution Analysis – Arbitrary Censoring.

This is a parametric analysis because a distribution will be assumed. It is arbitrary censoring because the data are not right-censored – "arbitrary" censoring covers the other options, including left-censored data. Two interval endpoints (see Chapter 3) must be specified for each observation. For nondetects the first variable ("Start") is set at zero, and the second variable ("End") is set to the detection limit. The procedure reads nondetects as being between zero and the detection limit (the "censoring interval"). Detected observations have the same value in both the Start and End variables. The resulting parameter estimates describe the distribution with the maximum likelihood of having produced a data set with the observed detected values and the proportions of censored data below each detection limit. Applying MLE to the Oahu data results in the output shown in Table 6.2.

If the software available to you does not include an option for "arbitrary" or "interval" or "left" censoring, right-censoring MLE may be used if a lognormal distribution is assumed. In that case the data are first log-transformed, then flipped by subtracting from a constant larger than the largest logarithm in the data set. Right-censored MLE software will use an upper bound of infinity for these flipped data. Because the data were first log-transformed, positive infinity will map to a value of 0 in original units. Location estimates (mean, median and percentiles) must be re-transformed (re-flipped and exponentiated) back into original units. Estimates of variability (standard deviation, variance, interquartile range) are the same before and after flipping and can be directly used (in log units). Assuming a normal distribution with right-censored MLE on flipped data will not produce correct results.

**Table 6.2.** Summary statistics for the Oahu data using the lognormal MLE method in Minitab®.

| | Estimate | Standard Error | 95.0% Normal CI Lower | Upper |
|---|---|---|---|---|
| Mean(MTTF) | 0.9453 | 0.1511 | 0.6910 | 1.2931 |
| Standard Deviation | 0.6559 | 0.1979 | 0.3631 | 1.1849 |
| Median | 0.7766 | 0.1325 | 0.5559 | 1.0850 |
| First Quartile(Q1) | 0.5088 | 0.1098 | 0.3334 | 0.7766 |
| Third Quartile(Q3) | 1.1854 | 0.1903 | 0.8653 | 1.6238 |
| Interquartile Range (IQR) | 0.6766 | 0.1472 | 0.4417 | 1.0363 |

Without the log transformation, the upper end of positive infinity for the flipped data will map to a lower end of negative infinity on the original scale. The MLE is being told that in the original scale, data values lower than zero are possible. Unless it is possible for the data to go below zero, location estimates will be too low when assuming a normal distribution for flipped data.

### *Interval censoring MLE*

In Chapter 3 the discussion of detection and quantitation methods included an example of a nonparametric ranking scheme that set ranks for observations measured below the detection limit as lower than ranks of data measured between the detection and quantitation limit. A similar procedure can be performed in a parametric framework by differentiating between the interval endpoints for the two types of data being input to maximum likelihood.

Consider the data set below, similar to but not identical with the Oahu data. A quantitation limit of 1 was used to censor results. Values measured between the detection limit of 0.5 and the quantitation limit were reported as qualified with a remark code signaling a warning. The remark code was dropped, as is often the case, and the numeric values used as though they were equivalent to values above the quantitation limit of 1. Values measured below 0.5 were reported as <1, an informative censoring procedure. The resulting dataset is given in the left-most column named "Values from the lab" below.

Biased (high) estimates of the mean and percentiles will result from informative censoring if the data are used as reported. To more correctly compute a mean for these data, maximum likelihood will be performed after re-coding censored values into intervals (StartInt and EndInt columns). All values reported as <1 (reported as below the quantitation limit though actually measured as below the detection limit) are considered to be within an interval between 0 and 0.5, showing that they were nondetects. This restores the original measured range of values and avoids informative censoring. To address uncertainty in the values between the two limits, all values measured between 0.5 and 1 are recoded to be within the interval 0.5 to 1 – between the limits. Quantified data, values measured above 1, are input with the same value in both the StartInt and EndInt columns. The software reads this as a quantified single value rather than an interval. Minitab® produces the summary statistics of Table 6.3 using MLE, assuming a lognormal distribution.

| Value from the lab | StartInt | EndInt | Start | End |
|---|---|---|---|---|
| <1 | 0 | 0.5 | 0 | 0.5 |
| <1 | 0 | 0.5 | 0 | 0.5 |
| <1 | 0 | 0.5 | 0 | 0.5 |
| <1 | 0 | 0.5 | 0 | 0.5 |
| <1 | 0 | 0.5 | 0 | 0.5 |
| <1 | 0 | 0.5 | 0 | 0.5 |
| <1 | 0 | 0.5 | 0 | 0.5 |
| <1 | 0 | 0.5 | 0 | 0.5 |
| <1 | 0 | 0.5 | 0 | 0.5 |
| <1 | 0 | 0.5 | 0 | 0.5 |
| 0.5 | 0.5 | 1.00 | 0.5 | 0.5 |
| 0.55 | 0.5 | 1.00 | 0.55 | 0.55 |
| 0.6 | 0.5 | 1.00 | 0.6 | 0.6 |
| 0.6 | 0.5 | 1.00 | 0.6 | 0.6 |
| 0.7 | 0.5 | 1.00 | 0.7 | 0.7 |
| 0.7 | 0.5 | 1.00 | 0.7 | 0.7 |
| 0.9 | 0.5 | 1.00 | 0.9 | 0.9 |
| 1.5 | 1.5 | 1.5 | 1.5 | 1.5 |
| 1.7 | 1.7 | 1.7 | 1.7 | 1.7 |
| 2.8 | 2.8 | 2.8 | 2.8 | 2.8 |
| 3.2 | 3.2 | 3.2 | 3.2 | 3.2 |
| 5.7 | 5.7 | 5.7 | 5.7 | 5.7 |
| 8.1 | 8.1 | 8.1 | 8.1 | 8.1 |

**Table 6.3.** Summary statistics from the lognormal MLE method for interval-censored data. Data below the detection limit and data between the limits were input as different intervals.

|  | Estimate | Standard Error | 95.0% Normal CI Lower | Upper |
|---|---|---|---|---|
| Mean (MTTF) | 1.35667 | 0.517576 | 0.642303 | 2.86557 |
| Standard Deviation | 2.81139 | 2.09520 | 0.652472 | 12.1138 |
| Median | 0.589618 | 0.184836 | 0.318957 | 1.08995 |
| First Quartile (Q1) | 0.246835 | 0.103307 | 0.108682 | 0.560605 |
| Third Quartile (Q3) | 1.40842 | 0.425117 | 0.779488 | 2.54482 |
| Interquartile Range (IQR) | 1.16158 | 0.383806 | 0.607862 | 2.21973 |

If informative censoring had been incorrectly used so that the 10 nondetects were assigned an interval between 0 and 1 instead of 0 to 0.5, a higher estimate of the mean would have been produced. This estimate would be biased high. Interval censoring methods for data analysis may alleviate the perceived need for informative censoring, resolving some of the conflict between a user's request for "numbers" and a laboratory analyst's protective reporting measures. Of course if the values measured between the limits of 0.5 and 1.0 are considered sufficiently reliable, those numbers can be used as individual values (Start and End values the same) rather than using an interval for data between the limits. The detection limit then becomes the reporting limit. MLE estimates become those of Table 6.4. Note there is little difference for parameter estimates using the two coding schemes. This should provide confidence that interval censoring captures most of the information present in the data between the limits.

## Nonparametric method: Kaplan-Meier

The standard method for estimating summary statistics of censored survival data is the nonparametric Kaplan-Meier (K-M) method. Yet as Ware and DeMets stated in 1976, "Although the Kaplan-Meier estimate is fundamental to survival data analysis, it is often overlooked when a left or right censored data [sic] arises in other settings" (Ware and DeMets, 1976). This has certainly been true in the setting of environmental sciences.

Kaplan-Meier is implemented in commercial statistics packages offering routines for survival analysis. However, it only accepts right-censored data. Left-censored data such as the Oahu arsenic data must be flipped in order for summary statistics to be estimated. Minitab® (version 14) has the K-M method in its

Stat > Reliability/Survival

menu, under a submenu selection named

Distribution Analysis – Right Censoring > Nonparametric Distribution Analysis.

**Table 6.4.** Summary statistics from the lognormal MLE method where only values below the detection limit are censored.

|  | Estimate | Standard Error | 95.0% Normal CI Lower | Upper |
|---|---|---|---|---|
| Mean (MTTF) | 1.31499 | 0.502001 | 0.622270 | 2.77888 |
| Standard Deviation | 2.74088 | 2.04200 | 0.636402 | 11.8045 |
| Median | 0.568820 | 0.178298 | 0.307726 | 1.05143 |
| First Quartile (Q1) | 0.237544 | 0.0993348 | 0.104663 | 0.539132 |
| Third Quartile (Q3) | 1.36208 | 0.411049 | 0.753931 | 2.46081 |
| Interquartile Range (IQR) | 1.12454 | 0.371099 | 0.588954 | 2.14719 |

Though left-censored data must first be flipped prior to K-M, for all nonparametric analyses the lower bound of zero is fortunately not of concern. The lowest values are considered simply the lowest values, and when flipped become the highest flipped values. Information that there is a lower bound of zero is not used. So with nonparametric methods, left-censored data are flipped, the procedures computed, and the results re-transformed. The maximum value in the Oahu data is 3.2, so 5 is arbitrarily chosen as a value larger than the maximum. Let M = this flipping constant, here equal to 5. Right-censored data are constructed by subtracting all observations from M

$$Flip_i = M_i - x_i \qquad (6.5)$$

for all observations $x_i$. The result is stored in a column labeled Flip in Table 6.5.

The K-M method produces estimates of the survival probability function S for right-censored data. The survival function S is the probability that a data value T > y for any specific values y. If computed using flipped data, S = Prob (Flip > y), or Prob (M-x > y), or Prob (x < M-y). The latter expression shows that survival probabilities are also the cumulative distribution function of the original x data. Computation of the survival function for the Oahu data is illustrated in Table 6.5.

Table 6.5 has one row for each of the detected observations in the data set. The survival probabilities S are computed for each detected value. Using the flipped values, the detected observations ("failures" or "deaths" in survival analysis terminology) are ranked from small to large, accounting for the number of censored data in between each detected observation. For example, there are eight censored values for the Oahu data at <2.0, between the flipped values of 2.2 and 3.3 (see Table 6.5). The rank of the surrounding flipped observations therefore jumps from 2 to 11; K-M places each nondetect at its detection limit prior to ranking. The "number at risk"

**Table 6.5.** Computation of Kaplan-Meier survival probabilities for the Oahu data (n=24). Censored data are accounted for through the rank statistic r.

| As, ug/L (detects) | Flip | rank r | # at risk b b = (n−r+1) | # detects d (# failed) | Incremental Survival p = (b−d)/b | S Survival Prob. |
|---|---|---|---|---|---|---|
| 3.2 | 1.8 | 1 | 24 | 1 | 23/24 | 0.9583 |
| 2.8 | 2.2 | 2 | 23 | 1 | 22/23 | 0.9167 |
| 1.7 | 3.3 | 11 | 14 | 1 | 13/14 | 0.8512 |
| 1.5 | 3.5 | 12 | 13 | 1 | 12/13 | 0.7857 |
| 0.9 | 4.1 | 17 | 8 | 1 | 7/8 | 0.6875 |
| 0.7 | 4.3 | 19 | 6 | 2 | 4/6 | 0.4583 |
| 0.7 | 4.3 | 19 | 6 | 2 | 4/6 | 0.4583 |
| 0.6 | 4.4 | 21 | 4 | 1 | 3/4 | 0.3438 |
| 0.5 | 4.5 | 22 | 3 | 3 | 0/3 | 0.0000 |
| 0.5 | 4.5 | 22 | 3 | 3 | 0/3 | 0.0000 |
| 0.5 | 4.5 | 22 | 3 | 3 | 0/3 | 0.0000 |

b equals the number of observations, both detected and censored, at and below each detected concentration. The number of detected observations at that concentration is d, where d is greater than 1 for tied values. The incremental survival probability is the probability of "surviving" to the next lowest detected concentration, given the number of data at and below that concentration, or $\frac{b-d}{b}$. The survival function probability is the product of the j = 1 to k incremental probabilities to that point, going from high to low concentration for the k detected observations.

$$S = \prod_{j=1}^{k} \frac{b_j - d_j}{b_j} \qquad (6.6)$$

For example, the survival function probability of 0.6875 for the concentration at 0.9 equals 0.7857 • (7/8). Note that for the case of ties, K-M assigns the smallest rank possible to each observation, rather than the average rank as is done for most nonparametric tests. K-M will assign a probability of 0 to the smallest observation (largest flipped value), if there are no nondetects below this value in the data set. This represents a plotting position of i/n for the empirical distribution function of flipped values, so that the probability of exceeding the last value is 0. If the smallest concentration is a censored value, as is usually the case, the smallest detected observation will have a nonzero exceedance probability, while probabilities are indeterminate for all nondetects below the lowest detected observation.

A plot of the survival function for the Oahu data is shown in Figure 6.3. The K-M analysis of Flip produces the following estimates of summary statistics:

| Mean (MTTF) | Standard Error | 95.0% Normal CI | |
|---|---|---|---|
| | | Lower | Upper |
| 4.0510 | 0.1647 | 3.7283 | 4.3738 |

| | | | |
|---|---|---|---|
| Median = | 4.3000 | | |
| IQR = | 0.4000 | Q1 = 4.1000 | Q3 = 4.5000 |

Location estimates for flipped data (mean, median, other percentiles) must be retransformed back into the original scale by subtraction from the constant M used to flip the data. When the above estimates for mean, median and Q1 and Q3 are subtracted from the flipping constant of 5, the results are those printed in Table 6.6. Estimates of variability (variance, standard deviation, standard error, IQR) are the same for both the flipped and original units; no retransformation is needed.

**Table 6.6.** Summary statistics using Kaplan-Meier for the Oahu arsenic data

| MEAN | STD DEV | Q1 | MEDIAN | Q3 |
|---|---|---|---|---|
| 0.949 | 0.807 | 0.500 | 0.700 | 0.900 |

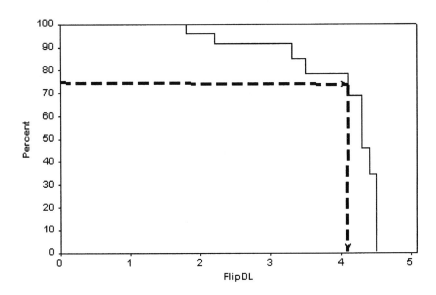

**Figure 6.3.** Survival function plot (Kaplan-Meier method) for the flipped Oahu data.

How were these summary statistics estimated by K-M? For percentiles, the estimate is the minimum X value on the survival function graph that is intersected by the line drawn at the probability value from the Y-axis. It is the smallest flipped observation having a survival probability equal to or less than the stated probability of the percentile. The $25^{th}$ percentile (Q1) has a survival probability (probability of exceedance) of 0.75. A horizontal line drawn from 0.75 on the Y-axis intersects the vertical line at an X-value of 4.1. Looking at Table 6.5, the flipped observation at 4.1 is the smallest flipped value for which the survival probability is 0.75 or less. Subtracting this from the flipping constant of 5, the $75^{th}$ percentile of the original data is 0.9. The process is similar for other percentiles. When more than 50% of data are censored, and the smallest observation (largest flipped value) is censored, the median cannot be estimated using K-M. A method which assumes some sort of model for the data distribution must be employed if an estimate for the median is required. Two possible methods for doing so include the fully parametric MLE (previous section) and the robust ROS (next section).

K-M methods include an estimate for the standard error of the survival function. Like the function S itself, the standard error is a step function that changes for each detected (uncensored) observation. Standard errors are computed most often to estimate confidence intervals around the estimated percentiles, describing the certainty with which that percentile value is known (see Chapter 7). Plots of survival functions often include the interval boundaries based on the standard error. The standard error formula, known as Greenwood's formula, is derived in many books on survival analysis, including Collett (2003),

$$\text{Std Error of S} = \text{s.e.}[S] = S \bullet \sqrt{\sum_{j=1}^{k} \frac{d_j}{b_j(b_j - d_j)}} \qquad (6.7)$$

where $b_j$ is the "number at risk" and $d_j$ is the number of detects (see Table 6.5) at each of the k values for detected observations.

The mean is generally considered less useful than the median in survival analysis, as distributions of medical or other 'lifetime' data are sufficiently skewed that the mean is not a typical value, but is strongly influenced by a few unusual values. The mean is computed by integrating the area under the K-M survival curve. To see why this is so, consider the usual equation for the mean of n observations,

$$\mu = \sum \frac{x}{n}.$$

When there are several observations at the same value, the equation can be stated as

$\mu = \sum \frac{f_i}{n} x_i$, where $f_i$ is the number of observations at each of the i unique values

of x, and $\frac{f_i}{n}$ is the proportion of the data set at that value. The mean is the sum of

the products of the proportion of data for each value times the magnitude of the observation's value. This is just what is accomplished when integrating under the K-M survival curve (Figure 6.4). The curve is divided by drawing horizontal lines at the value of each detected observation. The resulting set of rectangles have as their height the estimated proportion of data at that value, with the proportions summing to 1. The width of the rectangle is the magnitude of the observation, x. The mean is estimated by multiplying the width of each rectangle by its height to get the area, and then summing over all rectangles.

When the largest flipped value (smallest observation on the original scale) is censored, the censoring level (flipped detection limit) is used as the upper bound of the integration, essentially using the detection limit as the value for the smallest observation. This produces an estimated K-M mean that is biased low (biased high in the original units) because the true value of the last observation is somewhere beyond the last (smallest) observation. With the zero bound that is present for most environmental variables, this bias is not large. Note that this occurs only at the smallest detection limit, so that the positive bias in the K-M mean is less that of the simplistic substitution of the detection limit for each nondetect. However, most K-M software will warn of this bias, and the user should be aware that there is a small positive bias for the K-M mean in the typical case for environmental data where the smallest concentration is censored.

Estimates of standard deviation are even of less interest than the mean in traditional survival analysis due to the skewness found in most survival data. The variance and standard deviation are not resistant to skewness and outliers, and so provide a poor measure of the variability of data when those data are strongly skewed. Minitab® does not produce a value for the standard deviation with its K-M proce-

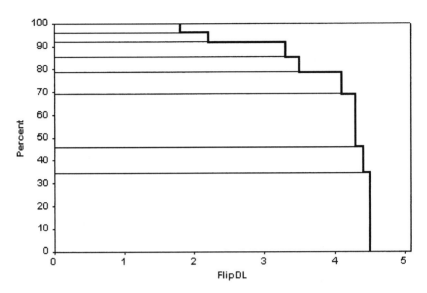

**Figure 6.4.** Computing the mean by integrating under the Kaplan-Meier survival curve. The total area is the K-M estimate of the mean.

dure, but it does return a value for the standard error of the mean. Although used primarily for constructing a confidence interval around the mean, the standard error of the mean can provide an estimate for the overall standard deviation "through the back door". For uncensored data, the standard error of the mean equals the standard deviation (s.d.) divided by the square root of the sample size n.

$$\text{std. error} = \frac{s.d.}{\sqrt{n}} \tag{6.8}$$

Therefore, multiplying the reported standard error by $\sqrt{n}$ will estimate the standard deviation. The standard deviation for the Oahu data estimated by this method is shown in Table 6.6.

### ROS: A "robust" method

Methods that calculate summary statistics with a regression equation on a probability plot are called "regression on order statistics," or ROS. One method of this type is fully parametric, and generally less efficient than maximum likelihood. In this fully parametric mode (described in the next paragraph) ROS has little advantage over MLE, and there is little reason to use it. However, a second implementation of ROS uses sample data whenever possible, assuming a distribution only for the censored portion of the distribution. Detected observations are used in their own right. In this robust form, abbreviated MR for Multiple-limit Regression by Helsel and

Cohn (1988), ROS is an attractive alternative to the more restrictive parametric assumptions of maximum likelihood. Its best application is to small data sets (n<30), where MLE estimation of parameters becomes inaccurate. The parametric and robust forms of ROS have often been confused in the literature. Travis and Land (1990) recommended the fully parametric method, though for justification they reference the results of Gilliom and Helsel (1986), who actually used the robust ROS. The two forms of ROS have different performance characteristics, and care should be taken to treat them separately.

ROS computes a linear regression for data or logarithms of data versus their normal scores, the coordinates found on a normal probability plot (e.g., see Figure 6.2). The regression parameters (slope and intercept) are computed using detected observations. Due to the definition of normal scores, fitting this line is fitting a normal distribution (a lognormal distribution if logs are used on the y-axis) to the observed data. For the fully parametric version of this method, the intercept and slope of the regression line estimate the mean and standard deviation, respectively, of the data or their logarithms. The intercept, the y value associated with a normal score of 0 at the center of the plot, estimates the mean of the distribution. The slope of the line equals the standard deviation, as normal scores are scaled to units of standard deviation.

If logarithms are used for the y-axis, the intercept and slope of a probability-plot regression estimate the mean and standard deviation of the logarithms. These summary statistics must be re-transformed to provide estimates of statistics in original units. Transforming moment statistics (mean and standard deviation) across scales with power transformations such as the logarithm results in transformation bias. The mean in one set of units will not provide an accurate estimate of the mean when converted to a second set of units using a power transformation. A simple example of transformation bias is given in Table 6.7. The mean logarithm of 2, re-transformed back to 100 in the original units, is biased low when compared to the mean of over 2000 for data in the original units.

Transformation bias is just as much of concern for parametric ROS as it was for MLE. If the estimates in log units are simply exponentiated, the resulting geometric mean estimates the median, and just as in Table 6.7 is biased low as an estimate of the mean. If equations 6.1 and 6.2 are used for re-transformation to avoid transformation bias, the resulting estimates are biased high (Cohn, 1988) for small samples due to the inaccuracy in estimating the standard deviation of the logarithms.

To avoid transformation bias, summary statistics can be computed using a robust approach to ROS. Using this approach, a more limited assumption of normality (or lognormality) is required – only that values below detection limits follow a specified distribution (Helsel and Cohn, 1988). The robust approach computes summary statistics in a different manner than fully-parametric ROS. After fitting the regression equation to detected observations on the probability plot, values for individual censored observations are predicted from the regression model based on their normal scores (the explanatory or x variable in the regression equation). Predicted values from the equation are used (or exponentiated and used if y is in log units) and

**Table 6.7.** An illustration of transformation bias.

| | Original Units | Logarithms (base 10) |
|---|---|---|
| | 1 | 0 |
| | 10 | 1 |
| | 100 | 2 |
| | 1000 | 3 |
| | 10000 | 4 |
| Mean: | 2222.2 | 2 |
| | $10^{\text{ mean log:}}$  100 | |

combined with detected observations to compute summary statistics as if no censoring had occurred. Retransforming individual values instead of the fitted parameters, and calculating summary statistics only after returning to the original units,
avoids the problem of transformation bias. Computations by robust ROS for the
Oahu arsenic data are given in Table 6.8 as an example.

Multiple detection limits are accounted for in the following way. First the probability of exceeding each detection limit is computed using the proportion of values
in the data set that are at or exceed each limit. For the Oahu data there are 2 out of
24 observations at or above the highest detection limit of 2 μg/L. So the probability of detection at 2 μg/L is 2/24, or 0.083. Then the probability of detection at the
next highest limit (1 μg/L) is computed. There are 14 observations below 2 μg/L
which can be compared to a detection limit of 1 μg/L. Of these, 2 observations are
at or exceed 1 μg/L and 12 observations are below 1 μg/L. So an estimate of the
probability of detection at a limit of 1 μg/L is (2/14)•0.917 + 0.083 = 0.214, where
0.917 is the probability of being at or below 2 μg/L (or 1 − 0.083). Finally, to determine the probability of detection at the lowest detection limit of 0.9 μg/L, of the 8
observations below 1 μg/L there is one at or exceeding 0.9 μg/L. So an estimate of
the probability of detection at a limit of 0.9 μg/L is (1/8)•0.786 + 0.214, or 0.313.

In general, the probability of exceeding the jth detection limit is

$$pe_j = pe_{j+1} + \frac{A_j}{A_j + B_j}\left[1 - pe_{j+1}\right] \tag{6.9}$$

where

$A_j$ = the number of observations detected between the $j^{th}$ and $(j+1)^{th}$ detection
limits, and

$B_j$ = the number of observations, censored and uncensored, below the $j^{th}$ detection limit.

When j = the highest detection limit, $pe_{j+1} = 0$ and $A_j + B_j = n$. The number of
nondetects below the $j^{th}$ detection limit is defined as $C_j$:

$$C_j = B_j - B_{j-1} - A_{j-1} \tag{6.10}$$

**Table 6.8.** Computation of summary statistics using robust ROS for the Oahu data (n=24).

| As, ug/L (detects) | log e Conc | Prob of Detection | Plot pos Percentile | rank r | Predicted logs | Observed + Estimated Concentration |
|---|---|---|---|---|---|---|
| 3.2 | 1.163 | | 0.972 | 24 | | 3.2 |
| 2.8 | 1.030 | | 0.945 | 23 | | 2.8 |
| <2 | | 0.083 | 0.815 | | 0.349 | 1.42 |
| <2 | | | 0.713 | | 0.134 | 1.14 |
| <2 | | | 0.611 | | −0.047 | 0.95 |
| <2 | | | 0.509 | | −0.215 | 0.81 |
| <2 | | | 0.407 | | −0.381 | 0.68 |
| <2 | | | 0.306 | | −0.559 | 0.57 |
| <2 | | | 0.204 | | −0.766 | 0.46 |
| <2 | | | 0.102 | | −1.05 | 0.35 |
| 1.7 | 0.531 | | 0.873 | 14 | | 1.7 |
| 1.5 | 0.405 | | 0.829 | 13 | | 1.5 |
| <1 | | 0.214 | 0.629 | | −0.018 | 0.98 |
| <1 | | | 0.471 | | −0.276 | 0.76 |
| <1 | | | 0.314 | | −0.543 | 0.58 |
| <1 | | | 0.157 | | −0.881 | 0.41 |
| 0.9 | −0.105 | | 0.737 | 8 | | 0.9 |
| <0.9 | | 0.313 | 0.344 | | −0.49 | 0.61 |
| 0.7 | −0.357 | | 0.589 | 6 | | 0.7 |
| 0.7 | −0.357 | | 0.491 | 6 | | 0.7 |
| 0.6 | −0.511 | | 0.393 | 4 | | 0.6 |
| 0.5 | −0.693 | | 0.295 | 3 | | 0.5 |
| 0.5 | −0.693 | | 0.196 | 3 | | 0.5 |
| 0.5 | −0.693 | | 0.098 | 3 | | 0.5 |

Plotting positions are then calculated in order to compute a normal score for each observation (see Table 6.8). Normal scores for detected observations are used to construct the regression equation relating the log of concentration to normal scores. Normal scores for nondetects are input to that regression equation to predict a log concentration, which is then retransformed to estimate concentrations for the set of nondetects. So plotting positions and normal scores are needed for both censored and uncensored observations. Plotting positions are at values spread equally between exceedance probabilities, and are computed separately for detected and nondetected observations. For the two detected observations above the highest detection limit of 2 µg/L, plotting positions are two values equispaced between (1−

$0.083 = 0.917$) and 1.0, or at $0.917 + \frac{1}{3} \bullet 0.083$ and $0.917 + \frac{2}{3} \bullet 0.083 = 0.972$. The

$C_3 = 8$ nondetects known to be <2 µg/L are spread evenly between probabilities of

0 and 0.917, or $(i/9) \bullet 0.917$, where $9 = C_3 + 1$ and i = 1 to 8, The two detected val-

ues between the detection limits of 1 and 2 g/L are spread evenly at 1/3 and 2/3 the distance between probabilities of $(1-0.214) = 0.786$ and 0.917. The $C_2 = 4$ nondetects known to be <1 µg/L are spread evenly between probabilities of 0 and 0.786, or at $(i/5)•0.786$, where $5 = C_2 +1$ and $i = 1$ to 4. The one detected observation between a <0.9 and 1 µg/L plots at a position halfway between detection probabilities of $(1-0.313) = 0.687$ and 0.786. The $C_1 = 1$ nondetect at <0.9 plots halfway between probabilities of 0 and 0.687. Finally, the six detected values below 0.9 plot at probabilities of $0+(i/7)•0.687$, where $i = 1$ to 6.

In general, plotting positions for detected observations are

$$pd_i = (1 - pe_j) + \left[\frac{i}{A_j +1}\right] • [pe_j - pe_{j+1}] \text{ for i=1 to } A_j \quad (6.11)$$

and for censored observations are

$$pc_i = \left[\frac{i}{C_j +1}\right] • [1 - pe_j] \text{ for i = 1 to } C_j \quad (6.12)$$

These equations follow the pattern of Hirsch and Stedinger (1987), who extended the traditional use of probability plotting in flood hydrology to the case of censored records of historical floods. After considering Bayesian and other methods for assigning plotting positions, their Appendix C provides equations 6.8 and 6.10 for determining plotting positions with multiple censoring levels. Helsel and Cohn (1988) extended these to equation 6.11 for censored data.

Estimated values produced for censored observations in the right-most column of Table 6.8 should not be assigned to any individual sample. For data sets with multiple observations below the same detection limit, there is no valid way to do so. Which estimate belongs to which sample is completely unknown. The corporate collection of estimates below each detection limit is sufficient to compute overall statistics, and yet does not allow the scientist to fall into the trap of indicating that the value for an individual censored observation is known, as is implied with simple substitution methods. Even when there is only one nondetect below a detection limit, declaring that a value is known for it is untrue. The value is known only to be within the interval from zero to the detection limit.

For the Oahu arsenic data, robust ROS estimates assuming a lognormal distribution are given in Table 6.9.

Shumway et al. (2002) have improved the robust ROS method by determining whether data best fit a lognormal, normal or square root–normal distribution prior to performing ROS. This was done by choosing the units that produced the largest log-likelihood statistics when fit by MLE. They state that one of these three distri-

Table 6.9. Summary statistics using robust ROS for the Oahu arsenic data.

| MEAN | STD DEV | Pct25 | MEDIAN | Pct75 |
|---|---|---|---|---|
| 0.972 | 0.718 | 0.518 | 0.700 | 1.103 |

butions generally matches the observed shape of environmental data. With this prior evaluation of distributional shape, they found that robust ROS produced estimates of the same quality as did MLE for moderate (n=50) sized data sets, and of better quality than MLE for small (n=20) data sets.

A side-by-side comparison of the estimates computed in this chapter for the Oahu arsenic data is given in Table 6.10. Results from using substitution were quite variable, and are known to be both arbitrary and inexact. There is no reason other than convenience to use them. The Kaplan-Meier method (K-M), independent of any distributional assumption, is the most generally applicable of the methods. If data appear to follow a lognormal distribution (the Oahu data are reasonably close) and there is a "large" set of uncensored data to adequately estimate parameters for the distribution (there is not here), then MLE should provide the best estimates. The value for "large" increases with increased skewness, but is generally at least 30 detected observations. These conditions are not often met in environmental studies. The ROS method, less dependent on a distributional assumption than MLE, is most useful for smaller data sets when MLE does not perform well. It can be an alternative to Kaplan-Meier when more than 50% of the data are censored and an estimate of the median is desired. Note that all three methods below the line in Table 6.10 result in more consistent estimates for this data set than do the substitution methods above the line.

## A review of comparison studies

Summary statistics for censored data is the most studied topic in the treatment of nondetects. The confusing element is that each article seems to find a different method to be 'best'. Why do conclusions differ so much on the choice of methods? Four important characteristics that strongly influence findings have varied among these studies. They are:

1. Sample size. MLE methods work far better for larger sample sizes (around 50) than smaller sizes. Some studies have used small samples, others larger.
2. Transformation bias. Some studies have computed estimates assuming a normal distribution, and simply stated that the results apply to lognormal or other distributions after transformation. This ignores transformation bias, the additional error resulting from moment statistics (mean and standard deviation) not being invariant to scale changes.
3. Robustness. Some studies generate data only from the same distribution that will be assumed in computing parametric methods. Method errors then reflect only the best-case scenario for parametric methods, ignoring the errors involved when the underlying distributions of data are mis-specified.
4. Details of method computation. For ROS, some studies use the fully-parametric approach, others the robust approach. For MLE, some studies use Cohen's table lookup, others the more-exact computer optimization (sometimes called the "expectation maximum" or "EM algorithm").

**Table 6.10.** Summary statistics using several estimation methods – Oahu arsenic data.

| METHOD | MEAN | STD DEV | Pct25 | MEDIAN | Pct75 |
|--------|------|---------|-------|--------|-------|
| Zero* | 0.567 | 0.895 | 0.000 | 0.000 | 0.700 |
| 1/2 dl* | 1.002 | 0.699 | 0.500 | 0.950 | 1.000 |
| dl* | 1.438 | 0.761 | 0.750 | 1.250 | 2.000 |
| Cohen's(ln)** | 1.034 | 0.882 | | | |
| MLE(ln) | 0.945 | 0.656 | 0.509 | 0.777 | 1.185 |
| robust ROS(ln) | 0.972 | 0.718 | 0.518 | 0.700 | 1.103 |
| K-M | 0.949 | 0.807 | 0.500 | 0.700 | 0.900 |

\* Methods are arbitrary and are not recommended for use.
\*\* Method is approximate and so not recommended for use.

The twelve papers reviewed below have directly compared methods for computing summary statistics using simulated data, to evaluate their performances. Root mean squared error (RMSE) and bias are usually computed as measures of the inadequacy of each method, with the best methods having low values for both. While this list of papers is not exhaustive, it does provide a summary of the major findings on this topic in environmental statistics to date. Most of the studies generated data from a single distribution, usually the lognormal distribution – Gleit (1985) used a normal distribution – and so do not evaluate errors due to mis-specifying the distribution of the data. This gives a great advantage to the parametric MLE and ROS methods, at the expense of nonparametric methods. Gilliom and Helsel (1986), Helsel and Cohn (1988), Kroll and Stedinger (1996) and She (1997) compared methods where the data distribution was not the same as the distribution assumed by maximum likelihood or ROS, to evaluate the robustness of each method.

1. Owen and DeRouen (1980) estimated means for air contaminant data, finding that MLE methods had high errors, especially for data of small sample size and a large proportion of censored values. They generated data of n = 5 to 50 having some true zeros, and found that the delta estimator, which assigns zeros to all censored values while modeling uncensored data as a lognormal distribution, had less error than MLE. This success might be attributed (though they did not do so) to the negative bias of assuming zeros counteracting the positive transformation bias produced when using formula 6.1 to retransform the estimate of the mean, which each of their methods used.

2. Gilbert and Kinnison (1981) evaluated deleting censored values, substitution, Cohen's table lookup and the fully parametric ROS method to estimate statistics of radionuclide data. They assumed lognormal data, and recommended Cohen's method and ROS for larger data sets with less than 50% censoring. With more censoring, they gave up trying to produce reasonable estimates, instead substituting the detection limit and reporting an estimate known to be biased.

3. Gleit (1985) generated small (n = 5 to 15) data sets from normal distributions censored at one detection limit and found that MLE methods did not fare well for such small sample sizes. Even though the assumed distribution was correct, MLE had difficulty estimating parameters with such little information. Substitution methods also had high errors – substituting the detection limit performed better than one-half or zero substitution, though no reasons are evident. The consistently best-performing method of the ones tested was a "fill-in with expected values", where an initial estimate of the mean and standard deviation are guessed, and using order statistics as with the robust ROS method in this chapter, estimates for individual observations produced. Then the mean and standard deviation are recomputed, and those values used to produce a second set of parameter estimates until convergence of the mean and standard deviation is achieved. Unfortunately, no evaluation was made of how to retransform estimates if logarithms of the data were used. No evaluation of robustness was made if the data were not normally distributed. Yet Gleit's study sounded themes that echo through later simulations – substitution methods work poorly, and MLE methods work poorly for small sample sizes.

4. Gilliom and Helsel (1986) compared substitution, standard (not table-lookup) MLE and the robust ROS procedures for a variety of generating data shapes censored at one censoring level. Substitution methods worked poorly. MLE methods worked well when the distribution assumed by the method (lognormal) reasonably matched that of the data. This was true for all but gamma distributions with high standard deviation and skew. The robust ROS method performed better on these high-skew distributions than did MLE, and performed similarly on the other distributions, and so was judged best overall.

5. Helsel and Cohn (1988) extended the results of Gilliom and Helsel (1986) to more than one detection limit. Whenever data did not follow the distribution assumed by maximum likelihood, and particularly with small sample sizes, the robust ROS generally produced better estimates for the mean and standard deviation than did MLE. Percentile estimates generally had smaller errors using a bias-corrected MLE than any other method. They introduced multiply-censored plotting positions to the ROS method which previously had been used only for flood frequency analysis. They also corrected MLE for transformation bias using Cohn's (1988) method.

6. Shumway et al. (1989) compared variations in computing maximum likelihood estimates of the mean and confidence interval on the mean. Originating data were from normal, lognormal, and square-root normal distributions of sample sizes 20 and 50, all with small variance and skew compared to some environmental data. Estimates improved when first determining from sample data to use either untransformed data, or log or square-root transformations based on selecting the maximum log likelihood of the three MLE equations, rather than always assuming lognormal or normal data. Confidence intervals for the mean were smaller (better) when the delta method, an asymptotic opti-

mization procedure, was used instead of bootstrapping.

7. Haas and Scheff (1990) applied a type-II bias correction to estimates from Cohen's table lookup (even though censored environmental data are type-I) and found the result to be better than the parametric ROS method for small sample sizes (n = 12) generated from normal distributions. As the percent of censored data increased, a "restricted maximum likelihood" adjustment performed slightly better than Cohen's method. When applied to a lognormal data set, none of the methods worked well, presumably due to transformation bias.

8. Rao et al. (1991) applied Cohen's table method to data generated from a normal distribution, comparing it to several substitution methods such as $DL/\sqrt{2}$ and fill-in methods from normal and uniform distributions. Their goal was to estimate the mean and confidence interval on the mean for skewed air-quality data. A bootstrap method was also employed to produce confidence bounds for the uniform fill-in. The bootstrap and MLE methods consistently produced better estimates of mean and confidence intervals. They then applied MLE to the logs of air data, and retransformed using equation 6.1, finding that the MLE no longer produced acceptable values. As they didn't recognize this was due to transformation bias and not the MLE itself, it kept them from recommending the MLE for regular use.

9. El-Shaarawi and Esterby (1992) computed exact biases for common substitution methods, stating that there is no further need to use Monte Carlo studies to evaluate these methods. They used an example to illustrate the fully-parametric ROS and MLE methods, though re-transformation bias was ignored. Two problems were highlighted:

   a) The value of the detection limit is not used in computing estimates by ROS, and they stated that ROS is therefore more directly related to Type II than Type I censoring.

   b) No estimates of standard error are computed with ROS and therefore confidence limits cannot be directly constructed [this was later solved by Shumway et al. (2002)].

10. Kroll and Stedinger (1996) implemented a robust procedure not only for ROS, but also for MLE and a probability-weighted moments estimator. They showed that the procedure used for circumventing transformation bias with robust ROS – retransforming single estimates from log space and computing summary statistics in the original units – could be done just as easily with MLE. Their 'robust MLE' performed somewhat better than robust ROS (the third method was not as good as these two). The advantages cited by Helsel and Cohn (1988) for robust ROS over MLE were shown to be due to the 'robust' adaptation, avoiding the transformation bias inherent in equation 6.1 for highly skewed and/or smaller sample sizes. Their work clearly shows the importance of the 'robust' adaptation, and that the order of choice for MLE and ROS methods should be: robust MLE > robust ROS >> lognormal MLE > (fully parametric) lognormal ROS.

11. She (1997) compared the lognormal MLE, the (fully-parametric) lognormal ROS, Kaplan-Meier and one-half substitution methods on both lognormal data and data from a gamma distribution. The nonparametric Kaplan-Meier was consistently the best or close to the best method for data from both distributions. The MLE performed well for data from the lognormal distribution when the skew was low. For highly skewed distributions, the moderate sample size (n=21) resulted in poor parameter estimates using MLE, even when the assumed distribution matched that of the data generated. The fully parametric ROS performed no better, and usually worse, than MLE.

12. Shumway et al. (2002) found that the robust ROS method ("robust regression ROS") had smaller errors than standard MLE for data from lognormal distributions for sample sizes of 25 and 50. ROS had similar errors to MLE when the MLE was followed by a jackknife method to compensate for retransformation bias, rather than using equations 6.1 and 6.2. When the original data were from a gamma distribution, however, they found that MLE estimates had less error than ROS when first transformed to approximate normality, especially for sample sizes of 50. They found, as did Cohn (1988), that the theoretical lognormal retransformation using equation 6.1 overcompensates for transformation bias and itself produces biased estimates. A jackknife correction for bias performed far better.

Additional papers to those above could certainly be cited – Singh and Nocerino (2002), for example, compared MLE, substitution, and ROS methods. However, their goal was not to model environmental data, but to estimate underlying 'background' data, assuming a mixture of background and contaminated samples. So they looked for methods which somehow deleted or downweighted all high values.

### A recommended course of action

A recommended course of action that takes these dozen articles into account (though undoubtedly it would not be endorsed by all of the above researchers) is given below, both in text and as Table 6.11. The recommendation of the Kaplan-Meier method for data with up to 50% censoring follows its predominant use in other disciplines as well as its well-developed theory. K-M is the nonparametric maximum likelihood estimator for constructing the survival function (Klein and Moeschberger, 2003). It requires no assumption of a particular distributional shape. If there were no censoring, K-M produces the familiar sample estimates for mean and percentiles.

The cutoff at 50 percent censoring in Table 6.11 reflects the fact that Kaplan-Meier does not provide an estimate for the median (other than <DL) when there are more than 50 percent nondetects. Some distribution must be assumed in that case, at least for the censored portion of the distribution. The cutoff at a sample size of 50 reflects the inability of MLE to accurately estimate parameters with small samples. Several of the above studies found that estimation errors increase dramatically between 60% and 80% censoring, and that above 80% censoring any estimates are merely guesses. Therefore at 80% censoring and above, methods that

dichotomize the data into proportions of detect/nondetect should replace attempts to estimate the central location or spread of a censored data set.

---

### *For less than 50% censoring*

Compute Kaplan-Meier estimates, the standard procedure in other disciplines and one that does not depend on the assumption of a distributional shape.

### *For large sample sizes (≥50) and 50–80% censoring*

Use MLE after first transforming the data to approximate normality. Re-transformed estimates of mean and variance must be corrected for bias by a method other than equations 6.1 and 6.2, such as the 'robust MLE' of Kroll and Stedinger (1996) or by a jackknife correction (Shumway et al., 2002).

### *For smaller sample sizes (< 50) and 50–80% censoring*

Use the robust MLE or robust ROS after first transforming the data to approximate normality. One measure of goodness-of-fit to a normal distribution is to maximize the probability plot correlation coefficient (Helsel and Hirsch, 2002), measuring the linearity of data on the probability plot. Alternatively, maximize the (negative) log-likelihood statistic produced by maximum likelihood. Following the recommendation of Shumway et al. (2002), possible transformations might be limited to logarithm and square-root.

### *Above 80% censoring*

Report the proportions of data below or above thresholds such as the maximum detection limit, rather than estimating statistics that are unreliable. High percentiles such as the $90^{th}$ or $95^{th}$ may be able to be estimated for large data sets. Any other estimates will be highly dependent on whichever distribution the data are assumed to follow.

**Table 6.11.** Recommended methods for estimation of summary statistics.

| | Amount of available data | |
|---|---|---|
| **Percent Censored** | **< 50 observations** | **> 50 observations** |
| < 50% nondetects | Kaplan-Meier | Kaplan-Meier |
| 50 – 80% nondetects | robust MLE or ROS | Maximum Likelihood |
| > 80% nondetects | report only % above a meaningful threshold | may report high sample percentiles (90th, 95th) |

# Exercises

**6-1**    The copper data from the Alluvial Fan zone of Millard and Deverel (1988) is found in the data set MDCu+ (use either MDCu+.mtw or MDCu+.xls). One observation has been changed from the data in their article. The largest detection limit of <20 was altered to become a <21, larger than all of the detected observations reported (the largest detected observation is a 20). Compute Kaplan-Meier estimates of the mean and median for two situations, one with the <21 in the data set and a second with the <21 removed from the data set. Demonstrate from the results that a censored observation whose threshold is above the largest detected observation has zero information content and can always be discarded. Also demonstrate why this is so by computing plotting positions by the robust ROS method for these data.

**6-2**    The silver.mtw data set contains analyses from 56 laboratories for a quality control standard silver solution (Helsel and Cohn, 1988). There are twelve detection limits reported by the different labs. Produce a survival function plot for the silver data using Kaplan-Meier software. Also produce a censored probability plot using robust ROS and with lognormal MLE. The ROS method can be computed using the Minitab macro Cros.mac. Compare and contrast the three plots. Which better illustrates how well the data are fit by the assumed distribution? Which would you use to get a rough estimate for the percentiles of the distribution?

**6-3**    Estimate the mean, standard deviation, median, 25th, and 75th percentiles of the silver data using (lognormal) maximum likelihood estimation, Kaplan-Meier, and the robust ROS methods. Compare and contrast the three results. How must the K-M percentiles be re-scaled in order to compare them with those from the other methods? Based on the percent of data censored, the sample size and the fit to the distribution, which method would you choose to use?

# 7

# COMPUTING INTERVAL ESTIMATES

Confidence, prediction and tolerance intervals are often needed for data that include nondetects. Two-sided confidence intervals bracket the likely values for a parameter such as the mean. The likelihood is represented by a confidence coefficient, a statement about how likely it is that the process used resulted in an interval that contains the true mean. A "95% confidence interval around the mean" is a statement of belief that the unknown true mean, a target of the investigation, is contained with a 95% probability between the lower and upper ends of the interval. The truth of that probability will rest on some assumptions about the distribution of the data if the method used is a parametric interval. When the data do not fit the assumed distribution well, the truth of the probability associated with parametric intervals will be in doubt. A "95% confidence interval" for an ill-fitting data set may in fact have a much higher probability than 5% of not including the true mean within the interval. Or it may be so wide that it is of little use.

One-sided confidence bounds are of interest when the concern is whether values are too large, or too small, but not both. An upper 95% confidence bound on the mean (sometimes abbreviated UCL95) is a statement that the true mean should be below the UCL95 with a 95% probability. If the UCL95 is below the relevant regulatory limit, there is more than a 95% probability that the true mean of the data lies below that limit. If the UCL95 is greater than the limit, there is more than a 5% probability that the true mean exceeds the limit, even though the observed sample mean may be below the limit. Confidence bounds such as the UCL95 have at times been written into environmental regulations. Of course, our interest is how to compute these bounds when some of the data are nondetects.

Prediction intervals provide a range bracketing the likely values for one or more individual observations not currently in the data set. Prediction intervals are wider than confidence intervals for the same set of data and same confidence coefficient. Intervals can be computed to enclose the range likely to hold one new observation, or many new observations, with a specified confidence. As with confidence intervals, parametric prediction intervals rely on the assumption that data come from a specific distribution, often the normal distribution. For parametric prediction intervals both the data used to construct the interval and the new data that are to fall within the interval are assumed to originate from the same distribution. If data do not fit the assumed distribution, they should be transformed prior to constructing the interval. Otherwise the intervals may be too large, or the stated confidence of inclusion may overstate the probability actually attained by the interval.

Tolerance intervals bracket possible values for percentiles of the distribution, and so include within their range a specified proportion of observations. Sample percentiles can be computed for the observed data (e.g., the sample 90th percentile equals or exceeds 90% of the measured values), but as with all other sample statistics, these only estimate the true underlying population statistic. A tolerance interval puts boundaries on the location of the true population percentile. An upper 95% tolerance bound on the 90th percentile, for example, provides a limit beyond which there is 95% confidence that no more than 10% of all population values fall.

Hahn and Meeker (1991) provide detailed descriptions of these three types of intervals. In the following sections, calculation of each type of interval is illustrated when some proportion of the data are recorded as nondetects. Each of the three types of intervals can be computed using one of the general approaches for censored data discussed so far – substitution, maximum likelihood, or nonparametric methods. Use of the first approach, substitution, is strongly discouraged.

## Parametric intervals

Parametric confidence, prediction and tolerance intervals are built using estimates of the mean and standard deviation, along with an assumption that data follow a normal distribution. If the distribution of data does not follow this shape, estimates of the interval endpoints can be severely in error. Parametric two-sided intervals follow the general equation of

$$\bar{x} - ks, \quad \bar{x} + ks$$

where $\bar{x}$ is the mean, s is the standard deviation, and k is a constant that is a function of the sample size n, the two-sided confidence coefficient $(1-\alpha/2)$, and the type of interval desired (confidence, prediction or tolerance interval). One-sided bounds concentrate the possible error rate $\alpha$ onto one side of the mean, following the general equation of $\bar{x} + ks$ for an upper confidence bound, and $\bar{x} - ks$ for the lower bound. For these one-sided bounds, k is a function of the sample size n, the one-sided confidence coefficient $(1-\alpha)$, and the type of interval desired.

Estimates for the mean and standard deviation may be computed using any of the techniques discussed in Chapter 6. Better parameter estimates will produce better interval estimates. Therefore estimates of mean and standard deviation from maximum likelihood (for large samples) or from the robust ROS or MLE estimators should produce the best interval estimates when the shape of data follows a normal distribution.

### Validity of assuming a normal distribution

Parametric intervals discussed in this book require that the data (or transformed data) used to construct the interval were randomly sampled from a population possessing the shape of a normal distribution, the familiar "bell-shaped curve". In that case most of the data will be in the center, with outlying data symmetrically departing from the center to more and more infrequent values. The center is defined by the mean and median, both of which are at the same value. The probability of being more than two standard deviations above the mean is identical to the probability of being more than two standard deviations below it.

Most investigators have found air, water, soils, and tissue concentrations to be somewhat skewed, unlike a normal distribution. A lower bound of zero concentration prevents the distribution from looking symmetric – concentrations can only go down so far, and no further. Infrequently occurring observations – outliers – occur primarily on the high (right) side, and so data are right-skewed. Due to both right-skewness and variability that is often proportional to the concentration, logarithms (and less frequently, square roots) are more often nearly symmetric, and a normal distribution for these transformed data can be more reasonably invoked than for the original data prior to transformation.

Parametric prediction intervals for one or more observations, and tolerance intervals around a specified proportion of data, are highly sensitive to the assumption of normality. If the data are skewed and these intervals computed directly using untransformed data, the endpoints of the intervals will not reflect the desired confidence level. The interval lengths will be unrealistic, perhaps going negative in the lower direction, an impossible result. The process of attempting to summarize a skewed distribution with symmetric intervals may produce an interval that is noticeably too large. A more insidious problem is that the interval may not contain the desired result as frequently as indicated by the confidence level. A supposed 95% confidence level for a prediction interval may, in fact, have only a 60% probability of containing the next observation, as one example.

However if prediction and tolerance intervals are cursed with strict dependence on the normality assumption, they also are blessed with an easy solution. Once the data are transformed to approximate normality and interval endpoints constructed on transformed data, those endpoints can be directly re-transformed back to original units while preserving their meaning and purpose. If natural logarithms of concentration have a nearly normal distribution, a 95% upper tolerance bound computed on the 90th percentile of logarithms may be retransformed by exponentiating its value. The result is a 95% tolerance bound on the 90th percentile of concentration. Prior to constructing prediction or tolerance intervals the Box-Cox transformation series, also called the Ladder of Powers (Velleman and Hoaglin, 1981) should be used to find a suitable transformation to near-normality. Then construct the desired interval, apply the reverse transformation and the result is the desired interval in original units.

Confidence intervals whose target is estimation of the mean have a different blessing and a different curse. Their curse is that an interval around the mean of data transformed using a power or other nonlinear transformation cannot be directly re-transformed to produce a confidence interval around the mean in original units. This is because the mean of transformed data once retransformed does not estimate the mean of the original data. The mean and median of transformed data are identical when the transformed data are symmetric. Once retransformed the resulting value remains an estimate of the median in original units, but not of the mean. So the geometric mean, the mean of logarithms retransformed back to original units, estimates the median (if the log-transformed data were symmetric) rather than the mean. Therefore transformation does not help in constructing confidence intervals on the mean, as it does for prediction and tolerance intervals. Confidence intervals constructed around the mean of the logs when re-transformed become confidence intervals for the median and not the mean, assuming the logs were symmetric.

How then can a confidence interval on the mean of skewed data be reliably estimated? An often-invoked blessing is the Central Limit Theorem, the property of a mean that states that the variability in estimates of the mean follows a normal distribution under certain conditions, even when the underlying data do not. When data sets are 'large' and data 'not too skewed', a normal theory confidence interval can be directly computed without transformation (Hahn and Meeker, 1991). But how large must 'large' be, and how much skewness is allowed? Boos and Hughes-Oliver (2000) show that the sample size required to invoke the Theorem is a function of the type of interval and the severity of skewness. For two-sided confidence intervals based on the t-statistic built from data of moderate skewness (skewness coefficient = 1), somewhere around 30 observations is large enough. However for a one-sided interval such as the upper 95% confidence bound on the mean (UCL95), a skewness coefficient of 1 results in a requirement of about 126 observations! With smaller sample sizes and right-skewed data, an upper confidence bound computed using the t-statistic will most often be too small (Boos and Hughes-Oliver, 2000), undershooting the true value ("miss on the left"). Environmental data generally have more than a moderate amount of skew (with a skew coefficient > 1), so that a sample size of 50 or more is not an unreasonable requirement to invoke the Central Limit Theorem for two-sided intervals. For one-sided confidence bounds, sample size requirements are quite large, greater than 126. If the skewness coefficient is known or can be estimated from past data, approximate sample size requirements can be determined from equations in Boos and Hughes-Oliver (2000). Even when sample sizes are sufficient, the resulting t-statistic confidence intervals must be considered approximate rather than exact, with the approximation getting appreciably worse as the confidence coefficient increases – as $\alpha$ gets small (Hahn and Meeker, 1991, p. 65).

Without invoking the Central Limit Theorem, Land (1972) developed a procedure to translate confidence intervals in log units back into intervals around the mean in original units. Land's method is discussed in the section on maximum likelihood. However, for small sample sizes and skewed data the estimate of standard deviation in log units is often poor. A poor estimate of this standard deviation causes Land's method to produce a confidence interval that is too wide (Singh et al., 1997). Bootstrapping (Efron and Tibshirani, 1986) is a newer and more efficient method than Land's for constructing confidence intervals on the mean of skewed data. Bootstrapping is a nonparametric method discussed in later sections.

## Nonparametric intervals

Rather than computing intervals using parameter estimates, traditional nonparametric intervals are based on the values of one or more observations in the data set.

Observations are chosen to be interval endpoints by their positions in the data set, called their order statistics. First the data are ordered from low to high. Interval endpoints are chosen at specific order statistics based on the sample size n, the desired confidence coefficient $(1-\alpha)$ or $(1-\alpha/2)$, and the type of interval (confidence, prediction or tolerance) to be computed. The values of observations located at each endpoint define the shape and width of the nonparametric interval.

Nonparametric intervals do not depend on an assumption of a normal distribution for their validity. The shape of the interval will reflect the shape of the data set. The tradeoff for this flexibility or robustness is that nonparametric intervals will be wider than parametric intervals when data do follow the assumed distribution. The distributional assumption is another piece of information used to construct a parametric interval. That additional information shortens the interval length when the data follow the assumed distribution. It will be misleading information producing misleading intervals when the data do not follow the distribution assumed by the process. In the latter case a nonparametric interval or an interval following appropriate transformation will provide better results.

A newer approach to computing nonparametric intervals is called bootstrapping. With bootstrapping the targeted statistic (mean, median, percentile, etc.) is repeatedly computed and the estimates stored. A thousand or more replications is typical. The collection of estimates approximates the distribution of the target statistic. The mean or median of estimates becomes the bootstrapped estimate at the center of the interval, and the appropriate low and high percentiles of the estimates forms the interval endpoints. For a 95% confidence interval, the $2.5^{th}$ and $97.5^{th}$ percentiles are used, leaving a total of 5% of the computed estimates outside the interval.

## Intervals for censored data by substitution

Substitution of an arbitrary constant, followed by computation of a parametric t-interval, is all too common. Intervals based on substitution will vary in quality. For example, Table 7.1 shows 95% confidence t-intervals for the mean of the Oahu data set. The three intervals are quite different, with the lower end of the DL interval being higher than the upper end of the Zero interval Yet there is no valid justification for arguing that one of these intervals is any better than another based on the data at hand. The primary problem with substitution is that the interval locations and widths are strongly influenced by the operating characteristics of the laboratory (the detection limits they settle on) and by the choice of an arbitrary fraction of that detection limit to substitute, rather than by the concentration of the target chemical that was in any given sample.

Substituting the possible extremes of zero and the detection limit will not necessarily produce intervals that bracket the range of possible interval widths. Though the mean varies monotonically as the substitution value changes, the standard deviation does not. For the Oahu data the standard deviation resulting from substitution of one-half DL was smaller than that when substituting zero or the detection limit. Interval widths of t-intervals using substitution follow the same pattern, and so will

**Table 7.1.** Confidence intervals (95%) for the mean of the Oahu data set using substitution.

| Substitution Method | Lower Limit | Upper Limit |
|:---:|:---:|:---:|
| Zero | 0.19 | 0.94 |
| One-half DL | 0.70 | 1.30 |
| DL | 1.12 | 1.76 |

not change monotonically. The maximum or minimum interval width may occur at an unknown substituted value somewhere between zero and the detection limit. So it is not possible to easily "bracket the extremes" of intervals using substitution.

Substitution should be avoided when computing interval estimates. There are better ways.

### Intervals for censored data by maximum likelihood

Interval estimates can be computed using maximum-likelihood estimates of the mean and standard deviation of a censored data distribution, placing these into the standard formulae for interval endpoints. The assumed distribution for the MLE is specified within the software. Example computations using Minitab® for many types of intervals follow. The data used are lead concentrations in the blood of herons in Virginia (Golden et al., 2003). The data are found in bloodlead.xls.

*Confidence interval for the mean (two-sided) assuming a normal distribution*

Assuming the data follow a normal distribution, upper and lower confidence limits for the mean are computed as

$$\bar{x} - t_{(1-\alpha/2)n-1} \bullet s/\sqrt{n}, \bar{x} + t_{(1-\alpha/2)n-1} \bullet s/\sqrt{n} \qquad (7.1)$$

where $t_{(1-\alpha/2)n}$ is the $1-\alpha/2$ th quantile of the t distribution, n is the sample size, and $s/\sqrt{n}$ is the standard error of the mean. If the interval is computed using data that are not normally distributed, the true probability of including the unknown population mean within this interval will be somewhat lower than $(1-\alpha\%)$. The dependence on the normality assumption can be relaxed as sample sizes increase. According to the Central Limit Theorem the probability of inclusion approaches $1-\alpha\%$ as n gets "large", where large increases as the skewness of data increases. For the skewness found in environmental data, "large" is often around 50–100 observations.

In Minitab® the mean, standard error, and confidence intervals are estimated and printed using maximum likelihood with the menu command:

Stat > Reliability/Survival > Distribution analysis (arbitrary censoring) > Parametric distribution analysis

Data are input in interval endpoint format with Blood0 as the start variable and

BloodPb as the end variable. The resulting output is:

| | Estimate | Standard Error | 95.0% Normal CI Lower | Upper |
|---|---|---|---|---|
| Mean (MTTF) | 0.0397452 | 0.0123321 | 0.0155748 | 0.0639156 |
| Standard Deviation | 0.0639343 | 0.0087089 | 0.0489537 | 0.0834990 |
| Median | 0.0397452 | 0.0123321 | 0.0155748 | 0.0639156 |
| First Quartile(Q1) | -0.0033778 | 0.0136753 | -0.0301810 | 0.0234253 |
| Third Quartile(Q3) | 0.0828682 | 0.0136438 | 0.0561268 | 0.109610 |
| Interquartile Range (IQR) | 0.0862460 | 0.0117481 | 0.0660376 | 0.112638 |

The 95% confidence interval on the mean extends from 0.016 to 0.064 µg/g lead. A normal probability plot (Figure 7.1) shows that the data are not normally distributed; they do not follow a straight line on the probability plot. The data set is of moderate size (27 observations). Therefore the probability that the population mean is somewhere within the 95% confidence interval is likely to be somewhat lower than 95%.

*Confidence bound for the mean (one-sided) assuming a normal distribution*
A one-sided confidence bound on the mean places the entire error probability $\alpha$ on either the upper or lower side. An upper confidence bound is computed using equation 7.2, assuming a normal distribution.

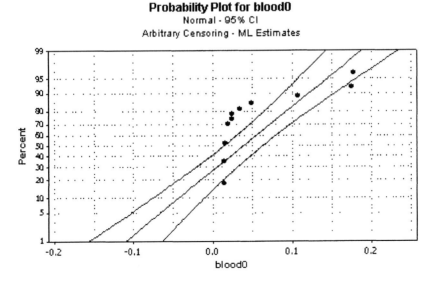

**Figure 7.1.** Probability plot of the censored blood lead data. Note the curvature, indicating non-normality.

$$\bar{x} + t_{(1-\alpha),n-1} \bullet s/\sqrt{n} \qquad\qquad (7.2)$$

A lower confidence bound is obtained by subtracting rather than adding from the estimate of the mean. A 95% upper confidence bound on the mean, assuming a normal distribution, is computed using censored MLE in Minitab® by selecting the "upper bound" option in the "Estimate" dialog box:

|  | Estimate | Standard Error | 95.0% Normal Upper Bound |
|---|---|---|---|
| Mean (MTTF) | 0.0397452 | 0.0123321 | 0.0600296 |
| Standard Deviation | 0.0639343 | 0.0087089 | 0.0799908 |
| Median | 0.0397452 | 0.0123321 | 0.0600296 |
| First Quartile (Q1) | -0.0033778 | 0.0136753 | 0.0191161 |
| Third Quartile (Q3) | 0.0828682 | 0.0136438 | 0.105310 |
| Interquartile Range (IQR) | 0.0862460 | 0.0117481 | 0.107906 |

The 95% upper confidence bound on the mean, assuming data follow a normal distribution, is 0.060.

*Confidence interval for the median assuming a normal distribution*
If the data are believed to follow a normal distribution the mean and median are the same. Therefore confidence intervals for the two are identical, as shown in the above Minitab® output. More typically, confidence intervals on the median are computed using the nonparametric process shown in a later section. The nonparametric intervals will have a true 95% probability of enclosing the population median, unlike parametric intervals when the data are skewed.

*Confidence bound for the median assuming a normal distribution*
A one-sided confidence bound for the median will be identical to that for the mean if data follow a normal distribution. The Minitab® results shown above for the one-sided upper bound on the mean have identical values in the row for the median.

*Confidence interval for a percentile assuming a normal distribution*
Confidence intervals around a percentile bracket the range of values within which the true population percentile is expected to be located, with $(1-\alpha)\%$ confidence. Assuming that data follow a normal distribution, the sample estimate of the pth percentile is:

$$\bar{x} + z_p \bullet s \qquad\qquad (7.3)$$

where is the pth percentile of the standard normal distribution. A two-sided confidence interval around a percentile larger than the median ($p > 0.5$) follows the formula:

$$\bar{x} + g'_{(\alpha/2)p,n} \bullet s, \quad \bar{x} + g'_{(1-\alpha/2)p,n} \bullet s \qquad (7.4)$$

where g' values are the percentiles of a non-central t distribution. The non-central t distribution is a function of the confidence coefficient ($1-\alpha$), the percentage p of the desired percentile, and the sample size n. Tables of the g' statistic for commonly-used values of $\alpha$ and p are found in Tables A.12a to A.12d of Hahn and Meeker (1991). Some statistical software, including Minitab®, will also produce values for the non-central t distribution.

For percentiles below the median ($p < 0.5$), a two-sided confidence interval is

$$\bar{x} - g'_{(1-\alpha/2)p,n} \bullet s, \quad \bar{x} - g'_{(\alpha/2)p,n} \bullet s \qquad (7.5)$$

Minitab® reports confidence intervals around the first quartile (25th percentile) and the third quartile (75th percentile) as part of its MLE distributional analysis. For other percentiles the intervals will need to be computed using equations 7.4 and 7.5.

The sample 90th percentile of the blood lead data is

$$0.0397 + 1.28 \bullet 0.064, \text{ or}$$
$$0.122$$

where 1.28 is the 90th percentile of a standard normal distribution ($p = 0.9$). The 95% confidence interval around this estimate for the 90th percentile is

$$(0.0397 + 0.846 \bullet 0.064, \quad 0.0397 + 1.932 \bullet 0.064), \text{ or}$$
$$(0.094, 0.163)$$

where 0.846 and 1.932 are the 2.5th and 97.5th percentiles of a non-central t-distribution from Tables A.12a and A.12d of Hahn and Meeker (1991) for $\alpha/2 = 0.025$, $p = 0.9$ and $n = 27$. Therefore the population 90th percentile of blood lead levels is between 0.094 and 0.163 µg/g with 95% confidence if the data follow a normal distribution.

*One-sided confidence bound for a percentile assuming a normal distribution*

One-sided confidence bounds for percentiles are computed in much the same way as the two-sided intervals. In environmental studies the objective for computing this one-sided bound is usually to define a limit that contains below (or above) it a specified proportion of the data. The equation for a one-sided upper confidence bound on a percentile is

$$\bar{x} + g'_{(1-\alpha)p,n} \bullet s \qquad (7.6)$$

where g' is the ($1-a$)•100th percentile of a non-central t distribution from Table A.12d in Hahn and Meeker (1991).

Suppose we want to determine a threshold that is higher than 90% of the blood lead levels in the population of herons. An upper 95% confidence bound on the $90^{th}$ percentile of lead concentrations will provide a threshold with only a 5% chance that the $90^{th}$ percentile of the population these data represent is higher than our estimate. This confidence bound for the heron blood lead data, assuming that they follow a normal distribution, is

$$0.0397 + 1.811 \cdot 0.064, \text{ or}$$
$$0.156$$

where 1.811 is $g'_{0.95, 0.9, 27}$ from Table A.12d of Hahn and Meeker (1991). We expect that 90% of all lead concentrations in the heron population represented by these birds would be below 0.156 µg/g, if the assumption of a normal distribution were reasonable. However, the data do not appear to follow a normal distribution, so this estimate is likely to be incorrect.

*Tolerance interval to contain a central proportion of the data*
Tolerance intervals bracket values containing a specified proportion of the data. Two-sided tolerance intervals are rarely used in environmental studies, perhaps because there are few applications that attempt to determine the location of a central proportion of data, with allowable exceedances at both high and low ends. Assuming the data follow a normal distribution, a two-sided tolerance interval follows the formula:

$$\bar{x} - g_{(1-\alpha)p,n} \cdot s, \quad \bar{x} + g_{(1-\alpha)p,n} \tag{7.7}$$

where tables of the g statistic (different than the g' statistic) are found in Table A10 of Hahn and Meeker (1991). Note that the tables are set up to use $\alpha$ rather than $\alpha/2$ for determining the g statistic for each end of the $(1-\alpha)\%$ confidence interval. The g statistic is also a function of the proportion p of the distribution to be included within the interval and the total sample size n. Neither Minitab® nor other standard statistical software computes tolerance intervals for a central proportion directly.

A tolerance interval expected to contain the central 90% of the blood lead data with 95% confidence is

$$(0.0397 - 2.184 \cdot 0.064, \quad 0.0397 + 2.184 \cdot 0.064), \text{ or}$$
$$(-0.100, 0.179)$$

where 2.184 is the g statistic from Table A.10a of Hahn and Meeker (1991) for $p = 0.9$ and $n = 27$. No more than five percent of the population of blood lead concentrations are expected to be less than the lower limit, and no more than five percent greater than the upper limit, with 95% confidence. The negative lower limit of the interval is a reminder that these data do not follow a normal distribution, and that a transformation should have been applied prior to computing this interval. This will be done in the later section on intervals for lognormal distributions.

*Tolerance bound for a proportion of the data*
    One-sided upper tolerance bounds estimate a value that exceeds p% of the pop-

ulation with $(1-\alpha)\%$ confidence. The percent of data (p) designed to be below the bound is often called the coverage. A one-sided tolerance bound is identical to the one-sided confidence limit for the equivalent (pth) percentile. So a 95% upper tolerance bound covering at least 90% of the data is identical to the 95% upper confidence bound for the $90^{th}$ percentile. Both are higher than p% of the data with $(1-\alpha)\%$ confidence. The equation for a one-sided tolerance bound with coverage p is the same as equation 7.6, above.

A 95% upper tolerance bound with 90% coverage of blood lead levels, assuming data follow a normal distribution, is

$$0.0397 + 1.811 \cdot 0.064, \quad \text{or} \quad 0.156$$

where 1.811 is the non-central t-statistic $g'_{0.95,0.9,27}$ from Table A.12d of Hahn and Meeker (1991). This is the same value obtained previously for the 95% upper confidence bound on the $90^{th}$ percentile. We would expect that at least 90% of blood lead concentrations in the population these data represent lie below 0.156, with 95% confidence, assuming these data follow a normal distribution.

*Prediction interval for one new observation, assuming a normal distribution*
Prediction intervals bracket the range of locations for one or more new observations not currently in the data set. Two-sided intervals are of interest if both extreme high and extreme low values of new observations are of concern. Obtaining a new observation beyond the limits of the prediction interval should happen only $\alpha\%$ of the time if nothing has changed and the new observation(s) come from the same distribution as did the existing data, in this case a normal distribution.

A two-sided prediction interval for normal distributions that covers the likely values for one new observation with $(1-\alpha)\%$ confidence is

$$\bar{x} - t_{(1-\alpha/2)n-1} \cdot \sqrt{1 + \tfrac{1}{n}} \cdot s, \quad \bar{x} + t_{(1-\alpha/2)n-1} \cdot \sqrt{1 + \tfrac{1}{n}} \qquad (7.8)$$

where t is from a Student's-t distribution with $n-1$ degrees of freedom. Note this is similar to the equation for the confidence interval around a mean, except that an additional term (a 1) appears under the square root sign. The uncertainty in prediction for a single new observation includes both the variability of the data (the standard deviation s) and the variability of the estimated mean (the standard error $\sqrt{1/n} \cdot s$ ). While the width of a confidence interval is determined only by the standard error, both terms contribute to the width of a prediction interval. Unlike a confidence interval, as sample sizes increase the width of a prediction interval goes no lower than $t \cdot s$.

A 95% prediction interval for the range of probable values for a new blood lead observation is

$$\left( 0.0397 - 2.056 \cdot \sqrt{1 + \tfrac{1}{27}} \cdot 0.064, 0.0397 + 2.056 \cdot \sqrt{1 + \tfrac{1}{27}} \cdot 0.064 \right), \text{ or}$$

$$(-0.094, \ 0.174)$$

where 2.056 is the $0.975^{th}$ quantile of a t-distribution with 26 degrees of freedom. The unrealistic negative lower end of the interval is a signal that these data do not fit a normal distribution well, and that a normal-theory prediction interval should not be used without prior transformation of the data.

*Prediction interval for several new observations, assuming a normal distribution*
An approximate prediction interval that covers the likely range of values for m new observations with $(1-\alpha)\%$ confidence is

$$\bar{x} - t_{1-\alpha/(2m),n-1} \bullet \sqrt{1+\tfrac{1}{n}} \bullet s, \bar{x} + t_{1-\alpha/(2m),n-1} \bullet \sqrt{1+\tfrac{1}{n}} \bullet s \qquad (7.9)$$

Including multiple predicted observations within the interval is accomplished by dividing $\alpha$ for the t-statistic by 2m rather than by 2. The t-statistic increases in value as m increases, widening the prediction interval. Tables for more exact and slightly smaller prediction intervals than those using equation 7.9 are found in Hahn and Meeker (1991).

A 95% prediction interval that should include values for 3 new blood lead observations is

$$\left( 0.0397 - 2.577 \bullet \sqrt{1+\tfrac{1}{27}} \bullet 0.064, 0.0397 + 2.577 \bullet \sqrt{1+\tfrac{1}{27}} \bullet 0.064 \right), \text{ or}$$

$$(-0.128, \ 0.208)$$

where 2.577 is the $0.992^{th}$ quantile $(1 - 0.05/6)$ of a t-distribution with 26 degrees of freedom.

Prediction intervals get very wide very quickly as m increases. Users often decide rather than to accept such wide intervals to use a tolerance interval instead. In this case the tolerance interval can be interpreted as giving the range of values that will include p% of all new observations, rather than all m observations, with $(1-\alpha)\%$ confidence.

*Prediction bound for new observations, assuming a normal distribution*
A one-sided prediction bound states the probable extreme in one direction for one or more new observations not currently in the data set. One example of its application is to determine a limit not likely to be exceeded by a new observation, based on an existing set of observations. For example, concentrations are measured in field blanks representing contamination due to the sampling and analytical processes. We might like to define a limit that, if exceeded, would indicate that the concentration in a new observation was greater than those in a blank. A one-sided 95% prediction interval would have only a 5% chance of being exceeded by a new observation similar to the blanks. That is a sufficiently small probability that an exceedance would be grounds for declaring that the concentration in a new observation was not simply due to contamination.

An upper prediction bound (assuming a normal distribution) that exceeds the likely values for m new observations, with $(1-\alpha)\%$ confidence, is

$$\bar{x} + t_{(1-\alpha/m)n-1} \bullet \sqrt{1+\frac{1}{n}} \bullet s \qquad (7.10)$$

A one-sided lower bound can be found by changing the plus sign following the mean in equation 7.10 to a minus sign.

For example, an upper prediction bound that should not be exceeded by one new blood lead observation with 95% confidence is

$$\left( 0.0397 + 1.706 \bullet \sqrt{1+\tfrac{1}{27}} \bullet 0.064 \right), \text{ or}$$
$$0.151$$

where 1.706 is the $0.95^{th}$ (1-0.05/1) quantile of a t-distribution with 26 degrees of freedom.

All of the above intervals, confidence, prediction, and tolerance, were for the situation where the data can be assumed to come from a normal distribution. We turn our attention now to the situation where the data do not appear to do so. First this situation is addressed by transforming the data to better fit a normal distribution, prior to using the equations already presented for a normal distribution. As a second approach, nonparametric intervals are created that require no assumption about the shape of the data distribution.

*The lognormal distribution*

The lognormal distribution is a skewed distribution of a variable x whose natural logarithms y = ln(x) follow a normal distribution. The lognormal distribution has been used by many investigators in environmental studies, as well as in other disciplines, to describe the shapes seen when a lower limit of zero is combined with the occurrence of infrequent yet annoyingly regular large outliers. The distribution has a flexible shape, appearing similar to a normal distribution when the skew is small (Figure 7.2, upper left), and much like an exponential decay function when skew is large (Figure 7.2, bottom right). To evaluate whether data follow a lognormal distribution the logarithms y are plotted on a normal probability plot, and the resulting pattern tested to see if it follows a straight line using the probability plot correlation coefficient, an analogue to the standard Shapiro-Wilks test for normality (Looney and Gulledge, 1985).

Tolerance and prediction intervals follow a simple process for lognormal distributions. The data are log-transformed, intervals are computed in the transformed units, and the interval endpoints are re-transformed back into the original units. The resulting intervals can be directly interpreted as prediction and tolerance intervals in the original units. Unfortunately, confidence intervals are not so simply computed.

Each of the intervals for a lognormal distribution uses estimates of the mean and standard deviation of the logarithms. In Minitab® these are calculated for censored data using MLE with the menu command

Stat > Reliability/Survival > Distribution analysis (arbitrary censoring) > Parametric distribution analysis

| Parameter | Estimate | Standard Error | 95.0% Normal CI Lower | Upper |
|---|---|---|---|---|
| Location | -4.26595 | 0.357922 | -4.96746 | -3.56443 |
| Scale | 1.40747 | 0.313299 | 0.909849 | 2.17728 |

Log-Likelihood = 7.548

Characteristics of Distribution

| | Estimate | Standard Error | 95.0% Normal CI Lower | Upper |
|---|---|---|---|---|
| Mean (MTTF) | 0.0377996 | 0.0153122 | 0.0170874 | 0.0836181 |
| Standard Deviation | 0.0944994 | 0.0786706 | 0.0184844 | 0.483118 |
| Median | 0.0140384 | 0.0050247 | 0.0069607 | 0.0283127 |
| First Quartile (Q1) | 0.0054329 | 0.0027094 | 0.0020443 | 0.0144385 |
| Third Quartile (Q3) | 0.0362750 | 0.0112877 | 0.0197124 | 0.0667537 |
| Interquartile Range (IQR) | 0.0308422 | 0.0100925 | 0.0162407 | 0.0585712 |

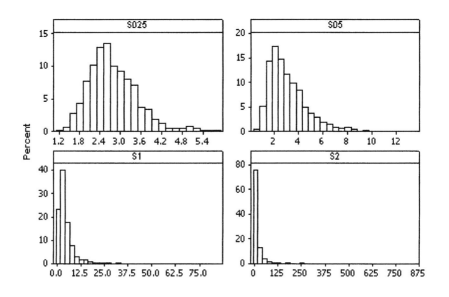

**Figure 7.2.** Histograms of four lognormal distributions with increasing skewness. Lowest skew at top left;  Highest skew at bottom right.

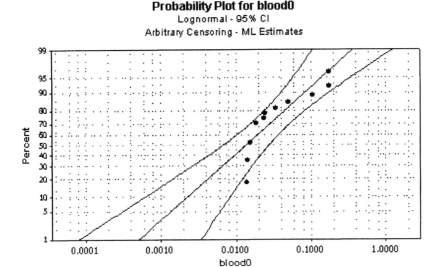

**Figure 7.3.** Probability plot of logarithms of the censored blood lead data.

while selecting lognormal as the fitted distribution. Results for the blood lead data
are plotted in Figure 7.3 and listed in the following output:

The mean and standard deviation of the (natural) logarithms are reported as the
location and scale parameter estimates, respectively. So the mean of the logarithms
$\bar{y}$ is estimated as –4.26 and the standard deviation of the logarithms $s_y$ as 1.407.
Estimates listed under "Characteristics of Distribution" are in the original units.
The mean (0.0377) was estimated by (same as equation 6.1)

$$\bar{x} = \exp(\bar{y} + s_y^2 / 2) \qquad (7.11)$$

and the standard deviation (0.094) as

$$s = \bar{x} \bullet \sqrt{\exp(s_y^2) - 1} \qquad (7.12)$$

The median (0.014) is the mean of the logarithms, retransformed

$$median_x = \exp(\bar{y}) \qquad (7.13)$$

Equations 7.11 and 7.12 are correct for large samples, but do not perform well
for small ($n < 50$) sample sizes. In particular, the estimate for the standard deviation
of the logarithms $s_y$ is inaccurately estimated by MLE for small samples. The result
is a biased estimate of the mean; its value is generally too large when data are right-
skewed. Various methods have been devised for correcting this bias (e.g., see

Shumway et al., 2002 and Cohn, 1988) but these are not implemented in standard statistical software.

*Confidence interval for the mean assuming a lognormal distribution*

If the data follow a lognormal distribution and sample size is large (about 100 or more), Land's method (Gilbert, 1987) can be used to calculate upper and lower confidence limits for the mean of lognormal data. Land's upper and lower two-sided confidence limits are computed as

$$\exp\left(\bar{y} + \frac{s_y^2}{2} + \frac{s_y \bullet H_\alpha}{\sqrt{n-1}}\right), \quad \exp\left(\bar{y} + \frac{s_y^2}{2} + \frac{s_y \bullet H_{1-\alpha}}{\sqrt{n-1}}\right) \qquad (7.14)$$

where tables of Land's H statistic are provided in an appendix by Gibbons and Coleman (2001). Land's H statistic is a function of both the desired confidence level $\alpha$ and of the standard deviation $s_y$ of natural logarithms. When sample sizes are smaller or when the coefficient of variation is greater than 1 (even for large sample sizes) these H-limits will not perform well (Singh et al., 1997). This includes most of the cases found in practice in environmental sciences, and therefore Land's method is rarely the best procedure. Singh et al. (1997) conclude that "for samples of size 30 or less, the H-statistic based UCL results in unacceptably high estimates of the threshold levels such as the background level contamination."

Gibbons and Coleman (2001) report several approximate methods for producing confidence intervals of lognormal data, but none are satisfactory. All of them either assume that the distribution of a statistic is normal when it is known not to be, or assume that the standard deviation of the logarithms is known, when in reality it is only estimated by $s_y$. This mis-specification of the standard deviation introduces large errors when sample sizes are small.

Currently, bootstrapping (Efron, 1981) is the most satisfactory method for computing confidence limits around the mean of lognormal data. Bootstrapping involves repeated computations of the same statistic thousands of times, each time on a temporary set of data chosen with replacement from the original data set. The mean of the computed estimates is the bootstrapped estimate of that statistic. Though there are several ways to compute a bootstrapped confidence interval, the method which assumes nothing about the distribution of the estimates is to take the 2.5[th] and 97.5[th] percentiles of the estimates as the 95% confidence interval endpoints. This was called the "percentile method" by its developer (Efron, 1981). Singh et al. (1997) strongly recommended the bootstrap or other nonparametric methods over Lands' H-statistic when computing confidence intervals for lognormal means.

Bootstrapped two-sided 95% confidence intervals for the mean of lognormal data can be computed by:

1. From the original set of n observations, sample with replacement to obtain a temporary set of n observations. Because some observations will be chosen more than once, the temporary set is usually not identical to the original data set.

2. Compute an MLE (or other method) estimate of the mean of the temporary set of data.
3. Save the estimate and repeat the process many times with new temporary sets of data. One thousand to ten thousand replicates is a commonly used range.
4. Compute the mean of the replicate estimates for the mean. This is the bootstrapped estimate of the mean.
5. Locate the 2.5th and 97.5th percentiles of the estimates of the mean. These are the endpoints of the two-sided 95% confidence interval for the mean.

Five percent of the estimates (2.5% on each side) are outside of the interval endpoints. The confidence interval may be asymmetric around the mean, reflecting that the distribution of the mean for small skewed data sets may not approach a normal distribution. More detail on bootstrapping is given in the section on nonparametric intervals.

*Confidence bound for the mean assuming a lognormal distribution*
    With large sample sizes (about 100 or more) and a coefficients of variation of 1 or less, Land's method (Gilbert, 1987) could be used to compute an upper confidence bound on the mean using equation 7.15.

$$\exp\left( \bar{y} + \frac{s_y^{\,2}}{2} + \frac{s_y \bullet H_{1-2\alpha}}{\sqrt{n-1}} \right) \qquad (7.15)$$

A lower confidence bound would obtained by substituting $H_{2\alpha}$ for $H_{1-2\alpha}$ in equation 7.15. The cautions of the previous section, including the comments by Singh (1997), apply equally well to one-sided bounds as to two-sided confidence intervals using Land's method for lognormal data.
    A more robust method for computing confidence bounds is bootstrapping. The one-sided upper 95% bound is the 0.95 quantile of the estimates of the means produced by the bootstrapping procedure. The lower 95% confidence bound is found at the 0.05 quantile of the bootstrapped means.

*Confidence interval for the median assuming a lognormal distribution*
    If the data follow a lognormal distribution the same central value is both the mean and median of the logarithms. The mean of the logarithms retransformed back into original units is the geometric mean, an estimate for the median of a lognormal distribution. A confidence interval around this median is calculated by retransforming the confidence interval for the mean of the logarithms back into original units.

$$x_{lo} = \exp\left( \bar{y} - t_{(1-\alpha/2)n-1} \bullet s_y / \sqrt{n} \right), x_{hi} = \exp\left( \bar{y} + t_{(1-\alpha/2)n-1} \bullet s_y / \sqrt{n} \right) \quad (7.16)$$

This interval has probability $(1-\alpha)$ of enclosing the population median as long as the data reasonably follow a lognormal distribution.

A confidence interval for the median blood lead concentration, assuming a log-normal distribution, is

$$x_{lo} = \exp(-4.266 - 2.056 \bullet 1.407/\sqrt{27}), x_{hi} = \exp(-4.266 + 2.056 \bullet 1.407/\sqrt{27})$$

or

$$(0.008 , 0.024)$$

where 2.056 is the t-statistic for $(1-\alpha/2) = 0.975$ and 26 degrees of freedom.

*Confidence bound for the median assuming a lognormal distribution*

An upper confidence bound around the median is computed by retransforming the upper confidence bound on the mean of the logarithms.

$$x_{hi} = \exp(\bar{y} + t_{(1-\alpha)n-1} \bullet s_y/\sqrt{n}) \tag{7.17}$$

The lower confidence bound substitutes a subtraction sign for the plus sign following $\bar{y}$. As before, the y values are logarithms and the x values are data in their original units.

An upper 95% confidence bound on the median blood lead concentrations is

$$x_{hi} = \exp(-4.266 + 1.706 \bullet 1.407/\sqrt{27}) , \quad \text{or}$$

$$0.022$$

where 1.706 is the t-statistic for $(1-\alpha) = 0.95$ and 26 degrees of freedom. The median blood lead concentration is expected to be no higher than 0.022 µg/g with 95% confidence if the data follow a lognormal distribution.

*Confidence interval for a percentile assuming a lognormal distribution*

Confidence intervals around a percentile of a lognormal distribution are computed in log units using the method for a normal distribution, and then retransformed back into original units. The sample estimate of the pth percentile is computed as:

$$x_p = \exp(\bar{y} + z_p \bullet s_y) \tag{7.18}$$

where $z_p$ is the pth percentile of the standard normal distribution.

A two-sided confidence interval around a percentile larger than the median $(p > 0.5)$ follows the formula:

$$\exp(\bar{y} + g'_{(\alpha/2)p,n} \bullet s_y, \bar{y} + g'_{(1-\alpha/2)p,n} \bullet s_y) \tag{7.19}$$

where tables of the g' statistic may be found in Tables A.12a to A.12d of Hahn and Meeker (1991). The g' statistic is based on a noncentral t distribution, and is a function of the confidence coefficient $(1-\alpha)$, the percentage p corresponding to the desired percentile, and the sample size n. For percentiles less than the median $(p < 0.5)$, the interval is

$$\exp(\bar{y} - g'_{(1-\alpha/2)p,n} \bullet s_y, \bar{y} - g'_{(\alpha/2)p,n} \bullet s_y) \tag{7.20}$$

For example, the sample $90^{th}$ percentile of the blood lead data, assuming the data are lognormal, is:

$$\exp(-4.266 + 1.28 \bullet 1.407), \text{ or}$$
$$0.085$$

where 1.28 is the $90^{th}$ percentile of the standard normal distribution (p = 0.9). The 95% confidence interval around this estimate for the $90^{th}$ percentile is

$$\exp(-4.266 + 0.846 \bullet 1.407, \quad -4.266 + 1.932 \bullet 1.407), \text{ or}$$
$$(0.046, 0.213)$$

where 0.846 and 1.932 are the g' statistics from Tables A.12a and A.12d of Hahn and Meeker (1991) for $\alpha/2 = 0.025$, p = 0.9 and n = 27. Therefore the $90^{th}$ percentile of the population of blood lead levels lies between 0.046 and 0.213 µg/g, with 95% confidence, if the data follow a lognormal distribution.

*Confidence bound for a percentile assuming a lognormal distribution*
One-sided confidence bounds for percentiles are computed in much the same way as the two-sided intervals. A confidence bound is appropriate if the objective is to define a limit that exceeds (or is lower than) a specified proportion (p) of the data, with a specified confidence $\alpha$. The equations for one-sided confidence bounds on a percentile are:

$$x_{hi} = \exp(\bar{y} + g'_{(1-\alpha)p,n} \bullet s_y) \quad \text{upper bound for p > 0.5} \qquad (7.21)$$
$$x_{lo} = \exp(\bar{y} - g'_{(1-\alpha)1-p,n} \bullet s_y) \quad \text{lower bound for p < 0.5}$$

where g' is from Tables A.12a to 12d in Hahn and Meeker (1991).
Suppose a threshold must be determined that is higher than 90% of the blood lead levels in herons represented by the sample data. An upper 95% confidence bound will reflect the uncertainty in the true $90^{th}$ percentile of lead concentrations. The confidence bound, assuming that the data follow a lognormal distribution, is:

$$\exp(-4.266 + 1.811 \bullet 1.407), \text{ or}$$
$$0.179$$

where 1.811 is $g'_{0.95,0.9,27}$ from Table A.12d of Hahn and Meeker (1991). We expect that at least 90% of all lead concentrations in blood of herons represented by this study will be below 0.179 µg/g, if the assumption of a lognormal distribution is reasonable for these data.

*Tolerance interval to contain a central proportion of lognormal data*
Two-sided tolerance intervals to contain a specified central proportion of the data, rarely used in environmental studies, provide allowable exceedances at both high and low ends. Assuming the data follow a lognormal distribution, a two-sided tolerance interval follows the formula:

$$\exp(\bar{y} - g_{(1-\alpha)p,n} \bullet s_y), \exp(\bar{y} + g_{(1-\alpha)p,n} \bullet s_y) \qquad (7.22)$$

where tables of the g statistic (different than the g' statistic) are found in Hahn and Meeker (1991). Note that the tables are set up to use $\alpha$ rather than $\alpha/2$ for determining the g statistic for each end of the $(1-\alpha)\%$ confidence interval. The g statistic is also a function of the proportion p of the distribution to be included within the interval and of the total sample size n. Neither Minitab® nor other standard statistical software computes tolerance intervals for a central proportion directly.

A tolerance interval expected to contain the central 90% of the heron blood lead data with 95% confidence is

$$\exp(-4.266 - 2.184 \cdot 1.407), \quad \exp(-4.266 + 2.184 \cdot 1.407), \quad \text{or}$$
$$(0.0006, 0.303)$$

where 2.184 is the g statistic from Table A.10 of Hahn and Meeker (1991) for p = 0.9 and n = 27. No more than five percent of blood lead concentrations in the population of herons are expected to be less than the lower limit of the interval, and no more than five percent above the upper limit, with 95% confidence. Unlike the two-sided tolerance interval for a normal distribution, the lower end of a lognormal interval is not negative. The lognormal distribution therefore is a more reasonable assumption than the normal distribution for these data. But perhaps a better fitting distribution than the lognormal could be found, improving the resulting intervals.

*Tolerance bound for a proportion of lognormal data*

One-sided upper tolerance bounds estimate a value that exceeds p% of the population values with $(1-\alpha)\%$ confidence. The percent of data (p) designed to be below the bound is often called the coverage. A one-sided tolerance bound is identical to a one-sided confidence limit for the equivalent (pth) percentile. So a 95% upper tolerance bound covering at least 90% of the data is identical to the 95% upper confidence bound for the 90th percentile. Both contain p% of the data with $(1-\alpha)\%$ confidence. The equation for a one-sided tolerance bound with coverage p is simply equation 7.21, above.

A 95% upper tolerance bound below which is at least 90% of blood lead levels, assuming that the data follow a lognormal distribution, is

$$\exp(-4.266 + 1.811 \cdot 1.407), \quad \text{or}$$
$$0.179$$

where 1.811 is $g'_{0.95,0.9,27}$ from Table A.12d of Hahn and Meeker (1991). This result is the same value obtained previously for the 95% upper confidence bound on the 90th percentile.

*Prediction interval for one new observation, assuming a lognormal distribution*

Prediction intervals bracket the range of locations for one or more new observations not currently in the data set. Two-sided intervals are of interest if both extreme high and extreme low values of new observations are of concern. Obtaining a new observation beyond the limits of the prediction interval should happen only $\alpha\%$ of the time if nothing has changed and the new observation(s) come from the same distribution as did the existing data.

A lognormal prediction interval that covers the likely values for one new observation with $(1-\alpha)\%$ confidence is

$$\exp\left(\bar{y} - t_{(1-\alpha/2)n-1} \bullet \sqrt{1 + \tfrac{1}{n}} \bullet s_y\right), \quad \exp\left(\bar{y} + t_{(1-\alpha/2)n-1} \bullet \sqrt{1 + \tfrac{1}{n}} \bullet s_y\right) \qquad (7.23)$$

where t is from a Student's-t distribution with $n-1$ degrees of freedom. See the discussion on prediction intervals in the normal distribution section for more detail.

A 95% prediction interval for the range of probable values for one new blood lead observation is

$$\left(\exp\left(-4.266 - 2.056 \bullet \sqrt{1 + \tfrac{1}{27}} \bullet 1.407\right), \quad \exp\left(-4.266 + 2.056 \bullet \sqrt{1 + \tfrac{1}{27}} \bullet 1.407\right)\right), \text{ or}$$

$$(0.0007, \ 0.267)$$

where 2.056 is the $0.975^{\text{th}}$ quantile of a t-distribution with 26 degrees of freedom. The lower end of a lognormal prediction interval will not go below zero, avoiding one of the primary problems for prediction intervals when assuming data follow a normal distribution.

*Prediction interval for several new observations, assuming a lognormal distribution*
An approximate prediction interval that covers the likely range of values for m new observations with $(1-\alpha)\%$ confidence is:

$$\bar{x} - t_{1-\alpha/(2m)n-1} \bullet \sqrt{1 + \tfrac{1}{n}} \bullet s, \bar{x} + t_{1-\alpha/(2m)n-1} \bullet \sqrt{1 + \tfrac{1}{n}} \bullet s \qquad (7.24)$$

assuming that data follow a lognormal distribution. Tables for more exact interval coefficients are found in Hahn and Meeker (1991).

A prediction interval with 95% confidence of including values for 3 new blood lead observations is

$$\left(\exp\left(-4.266 - 2.577 \bullet \sqrt{1 + \tfrac{1}{27}} \bullet 1.407\right), \quad \exp\left(-4.266 + 2.577 \bullet \sqrt{1 + \tfrac{1}{27}} \bullet 1.407\right)\right),$$

$$\text{or } (0.0003, \ 0.563)$$

where 2.577 is the $0.992^{\text{th}}$ quantile (1- 0.05/6) of a t-distribution with 26 degrees of freedom.

*Prediction bound for m new observations, assuming a lognormal distribution*
A one-sided prediction bound is used to determine a limit not to be exceeded by (or lower than) one or more new observations, based on an existing set of observations. Exceedance signifies that the new observation represents a different population than the existing data, with $(1-\alpha)\%$ confidence. See the section on normal distribution prediction bounds for more detail.

A one-sided upper prediction bound for a lognormal distribution that exceeds the likely values for m new observations, with $(1-\alpha)\%$ confidence is:

$$x_{hi} = \exp\left(\bar{y} + t_{(1-\alpha/m)n-1} \bullet \sqrt{1 + \tfrac{1}{n}} \bullet s_y\right) \qquad (7.25)$$

A one-sided lower bound $x_{lo}$ can be found by changing the plus sign to a minus sign following $\bar{y}$, the mean of the logarithms in equation 7.25.

An upper prediction bound that will likely not be exceeded by one new blood lead observation from the same population as the original data, with 95% confidence, is

$$\frac{\exp\left(-4.266 + 1.706 \bullet \sqrt{1 + \frac{1}{27}} \bullet 1.407\right)}{0.162}, \text{ or}$$

where 1.706 is the $0.95^{th}$ (1–0.05/1) quantile of a t-distribution with 26 degrees of freedom.

### Using other transformations

Other transformations may be used to construct intervals similar in purpose to those listed for the lognormal distribution, above. The general procedure is to:

1. Find a transformation that produces data close to a normal distribution.
2. Compute an interval on the transformed data.
3. For confidence intervals on percentiles, prediction intervals and tolerance intervals, retransform the interval endpoints directly back into original units. Confidence intervals on the mean and standard deviation are best performed using bootstrapping (see later section on bootstrapped intervals). Alternatively, complicated procedures based on the mathematics of the transformation itself may be possible, though the limited success with Land's method for small to moderate sized lognormal data sets should caution against using those type of methods.

Several methods can be used to determine which transformation best produces data that follow a normal distribution. Modern statistical software will easily produce a probability plot of the transformed data, comparing percentiles of the transformed data set to percentiles of a normal distribution. If the transformed data follow a normal distribution their points will plot on a straight line. In addition to choosing the units that visually produce the straightest data, there are several numerical measures for judging the adequacy of alternate transformations.

A. The probability plot correlation coefficient (PPCC – see Looney and Gulledge, 1985) increases to a value of 1.0 as data on the probability plot approach a straight line. Choosing the transformation that produces a PPCC closest to 1.0 is the measure most closely associated with the probability plot itself. The coefficient is also used in a test for normality.
B. Two other popular tests for normality are the Anderson-Darling (Stephens, 1974) and Shapiro-Wilk (Shapiro and Wilk, 1965) tests. In each case the null hypothesis is that the data follow a normal distribution. The transformation with the least-significant test statistic (largest p-value) would be produced from the transformation closest to a normal distribution by this measure.
C. After transforming data to approximate normality, compute the MLE for estimating mean and standard deviation assuming a normal distribution. The highest log-likelihood statistic will result from the transformation producing data closest to a normal distribution (Shumway et al., 2002).

Any of these criteria should lead to a reasonable transformation for the data. Several authors recommend limiting Box-Cox power transformations to values on the Ladder of Powers – transformations easy to interpret. Shumway et al. (2002) recommend considering only the log and square root transformations when computing summary statistics and interval estimates of environmental data, because these transformations mitigate the effects of the severity of right-skewness commonly seen in these data.

## Intervals using 'robust' parametric methods

Methods other than maximum likelihood may also be used to estimate the mean and standard deviation of data or of its transformed values. Equations from the previous sections are then used with these estimates to compute the intervals. Huybrechts et al. (2002) found that robust methods performed better than MLE for the small, skewed data sets common to environmental studies. Using either the robust ROS method described in Chapter 6, or the 'robust MLE' method described by Kroll and Stedinger (1996) should produce estimates for interval calculations that are as good as or better than MLE for small (< 50), skewed data sets.

## Nonparametric intervals for censored data

Nonparametric intervals assume no shape for the underlying data when computing locations of interval endpoints. The interval shape generally reflects the shape of the observed data, whatever that may be. The primary benefit of a nonparametric interval is that the probabilities of the targeted measure being within and outside the interval are correct regardless of the shape of the underlying data. A 95% nonparametric confidence interval for the median will have a 5% probability of not including the true value, whether the underlying data were normal, lognormal, or some other distributional shape.

The primary drawback of a nonparametric interval is that it makes no use of the information present in a distributional assumption. Therefore if the data do follow a known distribution, the width of a nonparametric interval will be larger than necessary – larger than the width of a parametric interval based on the correct data distribution at the same confidence level. The choice of whether to use a nonparametric interval should be made based on how uncertain the analyst is that the data follow a specific distribution. If the data do not fit the assumed distribution well, the parametric interval may be wider, and will be less accurate, than a nonparametric interval.

The primary method for computing nonparametric intervals is to order the data from smallest to largest, counting in from the ends a specific number of observations. The number of observations is determined by binomial probabilities. The ordered observations are called the "order statistics" of the data set, and for one detection limit are known as well as for any data that contain ties. For example, counting in approximately 20 observations from each end of a data set of 50 observations, selecting the 20th and 30th smallest observations, produces an approximate 90% confidence interval on the median. If the low end of the interval drops below the one detection limit, the low end may always be specified as "<DL" without making any unfounded assumptions or statements. For multiple detection limits a simple nonparametric interval may always be obtained after re-censoring all observations below the highest detection limit (HDL). Any interval endpoints below that threshold are called "<HDL". However, nonparametric intervals for multiply-censored data can be obtained with more precision by using methods based on Kaplan-Meier statistics rather than by censoring at the highest detection limit.

As the width of nonparametric intervals jumps from point to point, the associated confidence coefficients jump as well. The result is an interval with only approximately the same confidence level as what is desired. If a 95% confidence interval around the median is desired, the closest interval may either be a 96% interval when counting in 5 points from each end, or a 93% interval when counting in 6 points. A typical decision is to use the interval that has no more than an $\alpha\%$ error. So the 96% interval with a 4% error rate would be chosen. Alternatively some software will interpolate between the two sets of endpoints in order to provide a pseudo-95% interval. Jumping from one confidence coefficient to the next is most severe when sample sizes are small; for larger sample sizes a set of observations can usually be found that are quite close to the desired level of confidence.

*Nonparametric binomial confidence interval for the median*

Nonparametric interval estimates for the median and other percentiles can be computed using binomial probabilities. Interval endpoints are chosen using binomial tables where the proportion p in the table is the percentage corresponding to the targeted percentile at the center of the interval. For an interval surrounding the median, the binomial table is entered at p = 0.5. Binomial tables are also programmed into most statistical software.

A two-sided confidence interval on the median is computed using

$$[x_L+1, n-x_L] \tag{7.26}$$

where $x_L$ is the entry in a binomial table with p = 0.5 whose probability is closest to but not exceeding $\alpha/2$. The confidence interval endpoints are data values $x_L+1$ in from both ends of the list of ordered observations.

For the heron blood lead data there are 15 nondetects at <0.02 as well as 4 detected observations below 0.02, so the smallest 19 observations are all <0.02. The sample median of the 27 observations is the (n+1)/2 = 14th smallest observation, and so <0.02. A 95% confidence interval on this median is found using a binomial table with p = 0.5 (representing the median), $\alpha/2 = 0.025$ and n = 27. In Minitab® the binomial table is accessed by the command

    Calc > Probability distributions > Binomial

Tabled entries are obtained by specifying input values as shown in Figure 7.4.

The Minitab output is

```
Binomial with n = 27 and p = 0.5

x   P( X <= x )        x   P( X <= x )
7    0.0095786         8    0.0261195
```

Minitab reports two possible values for $x_L$, either 7 or 8, because neither produce an exactly 95% confidence interval. If 7 is used the resulting significance level will be 2•(0.0096) = 0.019, or a confidence level of 98.1%, larger than the desired level. If 8 is used the resulting significance level will be 2•(0.026) = 0.052, or a 94.8% confidence interval, slightly less than the desired 95% level. Most practitioners would choose the latter as being quite close to 95%. To obtain interval endpoints, count in 8+1 = 9 observations from each end of the ordered data set. These would be the 9th and (27-9) = 18th smallest observations, with values of [<0.02, <0.02]. Using binomial probabilities the nonparametric interval estimate states only that the median is somewhere below 0.02 with 95% confidence.

Minitab® also provides a nonlinear interpolation between these values to obtain an approximate 95% confidence interval, using the

Stat > Nonparametrics > 1–sample sign

procedure. This interval around the median is directly related to the sign test. Observations outside the $(1-\alpha)$% interval are sufficiently far from the center that the sign test would declare their value to be significantly different than the observed median at the $\alpha$% level. Values within the interval are not significantly different. The interpolated 95% confidence interval is given as [<0.02, 0.020], as shown in the output below.

Sign confidence interval for median

|           | N  | Median  | Achieved Confidence | Confidence Interval Lower | Upper   | Position |
|-----------|----|---------|---------------------|---------------------------|---------|----------|
| blood Pb  | 27 | 0.01999 | 0.9478              | 0.01999                   | 0.02000 | 9        |
|           |    |         | 0.9500              | 0.01999                   | 0.02009 | NLI      |
|           |    |         | 0.9808              | 0.01999                   | 0.02352 | 8        |

"NLI" stands for nonlinear interpolation. Both the values of 0.01999 and 0.02000 should be read as actually < 0.02 – the sign test does not read less-than signs. The upper interval endpoint of 0.02009 appears to be above the detection limit of 0.02, but is actually an interpolation between values of <0.02 and 0.02352, one of the detected observations. Whenever the interpolation is between a censored and uncensored observation, the reported value is not exactly known. A value of 0.02 was used as the low end of the interpolation, which is too high. Unless both interval endpoints are above the detection limit, using an interpolated interval remains inexact. Use the closest exact interval, here the 94.8% interval, rather than

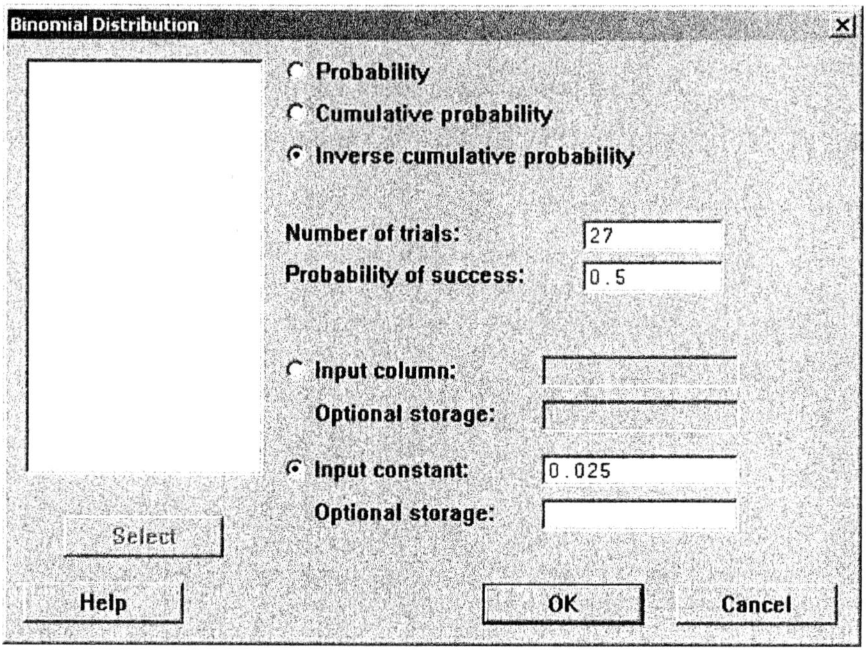

**Figure 7.4.** Minitab® dialog box to obtain the order statistics value for a 95% nonparametric confidence interval on the median (p = 0.5) of 27 blood lead observations.

the interpolation.

Nonparametric intervals based on binomial probabilities work reasonably well for one detection limit, but cannot be used for more than one limit except by censoring all values below the highest detection limit (HDL) as <HDL. If an upper bound is all that is desired, and binomial probabilities result in an upper bound that is higher than the HDL, all is fine. However, nonparametric intervals that account for multiple detection limits can also be produced using Kaplan-Meier statistics. K-M methods have the advantage of being the standard methods in medical statistics, and do not require re-censoring to the highest limit. K-M intervals are illustrated in the next section.

*Nonparametric confidence intervals based on Kaplan-Meier methods*

Kaplan-Meier methods estimate the survival function (or cdf or percentiles) of the data without assuming that data follow any particular distribution. K-M methods are coded in statistics software assuming that data will be right-censored. Left-censored environmental data must first be flipped into right-censored format (see Chapter 3) prior to computing the estimates. The blood lead data were flipped to a right-censored format by subtracting the lead concentrations from 1.0, a value larger than the maximum concentration. The flipped data are processed using the Kaplan-Meier procedure, invoked in Minitab® using

Stat > Reliability/Survival > Distribution analysis (right-censoring) >
Nonparametric Distribution Analysis

A plot of the survival function with confidence bands around the function is
shown in Figure 7.5.

The K-M table of survival probabilities and their 95% confidence intervals are
shown in Table 7.1. Added to the table is a column of re-transformed blood lead
values (Lead), computed as (1–Time) where Time represents the flipped data. The
Kaplan-Meier estimate for the pth percentile is the observation with the largest survival probability $\leq p$. For this example, the K-M estimate of the median is at the
observation with a survival probability of 0.35, the largest probability $\leq 0.5$. The
median is therefore at a Time of 0.984, or a lead concentration of $(1–0.984) = 0.016$.

The "95.0% Normal CI" K-M intervals are confidence intervals on survival
probabilities rather than on data values. They represent vertical distances on a survival plot such as Figure 7.5, rather than the horizontal interval that would provide
a confidence interval around the median or other percentiles.   Klein and
Moeschberger (2003) note that these probability confidence intervals ("linear confidence intervals") are quite inaccurate for "small" data sets of less than 200 observations!  See Klein and Moeschberger (2003) for more information if intervals
around survival probabilities are of interest.  Confidence intervals around the median or other percentiles are provided in the next sections.

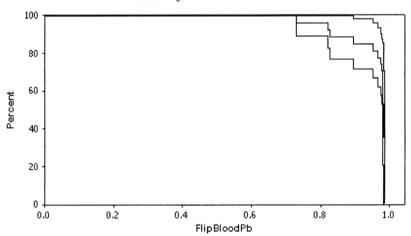

**Figure 7.5.**  Survival function for the flipped blood lead data.

**Table 7.1.** K-M table of survival probabilities and their 95% confidence intervals.

| Mean (MTTF) | Standard Error | 95.0% Normal CI Lower | Upper |
|---|---|---|---|
| 0.957310 | 0.0125752 | 0.932663 | 0.981957 |

Median = 0.984483
IQR = 0.0102426   Q1 = 0.975472   Q3 = 0.985714

Kaplan-Meier Estimates

| Time | Lead | Number at Risk | Number Failed | Survival Probability | Standard Error | 95.0% Normal CI Lower | Upper |
|---|---|---|---|---|---|---|---|
| 0.731034 | 0.261 | 27 | 1 | 0.962963 | 0.036345 | 0.891729 | 0.99999 |
| 0.822951 | 0.177 | 26 | 1 | 0.925926 | 0.050401 | 0.827142 | 0.99999 |
| 0.825926 | 0.174 | 25 | 1 | 0.888889 | 0.060481 | 0.770348 | 0.99999 |
| 0.893939 | 0.106 | 24 | 1 | 0.851852 | 0.068367 | 0.717854 | 0.98584 |
| 0.950847 | 0.049 | 23 | 1 | 0.814815 | 0.074757 | 0.668294 | 0.96133 |
| 0.966038 | 0.034 | 22 | 1 | 0.777778 | 0.080009 | 0.620963 | 0.93458 |
| 0.975472 | 0.025 | 21 | 1 | 0.740741 | 0.084337 | 0.575443 | 0.90603 |
| 0.976471 | 0.024 | 20 | 1 | 0.703704 | 0.087877 | 0.531468 | 0.87593 |
| 0.981356 | 0.019 | 4 | 1 | 0.527778 | 0.166001 | 0.202422 | 0.85312 |
| 0.984483 | 0.016 | 3 | 1 | 0.351852 | 0.181330 | 0.000000 | 0.70724 |
| 0.985714 | 0.014 | 2 | 1 | 0.175926 | 0.153932 | 0.000000 | 0.47762 |
| 0.986275 | 0.014 | 1 | 1 | 0.000000 | 0.000000 | 0.000000 | 0.000000 |

*Nonparametric confidence intervals for the median based on Kaplan-Meier*
   Three methods of computing intervals for the median of a multiply-censored variable are presented below. All avoid assumptions about the shape of the data distribution, though the first method assumes that the variation in estimates of a single survival probability is asymptotically normal. The second method is based on an adaptation of the sign test for multiply-censored data. It is fully nonparametric, but uses a large-sample normal approximation for the test statistic. The third method is bootstrapping, which makes no assumptions about the distribution of data or any test statistic. Bootstrapping is discussed in its own section. The bootstrap and sign-test methods provide reliable nonparametric estimates of confidence intervals for percentiles of data with nondetects.
   The first type of nonparametric interval based on Kaplan-Meier statistics uses Greenwood's formula for the standard error of the survival function, s.e.$[\hat{S}]$ (equation 6.6), as the measure of the vertical variability around the survival curve. Greenwood's standard error is usually printed in output from K-M software as a column of standard errors, one value for each survival probability. The standard error for a percentile, the variability in a horizontal direction on a plot of the survival curve, is related to Greenwood's standard error by estimating the probability density function or pdf at that percentile (Collett, 2003).
   The standard error of the median is related to Greenwood's standard error s.e. $[\hat{S}]$ by equation 7.27 (Collett, 2003).

$$s.e.[median] = \frac{s.e.[\hat{S}]}{pdf[median]} \tag{7.27}$$

where pdf[median] is the probability density function evaluated at the median. The pdf is approximated using equation 7.28 (Collett, 2003).

$$pdf[median] = \frac{\hat{S}(T^+(0.55)) - \hat{S}(T^-(0.45))}{T^-(0.45) - T^+(0.55)} \tag{7.28}$$

where $T^+(0.55)$ is the largest survival time whose estimated survival probability exceeds 0.55, $T^-(0.45)$ is the smallest survival time whose estimated survival probability is less than or equal to 0.45, and $\hat{S}$ is the estimated survival probability. The $(1-\alpha)\%$ confidence interval around the sample median $\hat{T}(0.5)$ is then

$$\hat{T}(0.5) - z_{1-\alpha/2} \bullet se[median], \quad \hat{T}(0.5) + z_{1-\alpha/2} \bullet se[median] \tag{7.29}$$

   As an example, the 95% confidence interval on the median blood lead concentration computed with equations 7.27–7.29 is:

$$pdf[median] = \frac{\hat{S}(T^+(0.55)) - \hat{S}(T^-(0.45))}{T^-(0.45) - T^+(0.55)} = \frac{0.704 - 0.352}{0.984 - 0.976} = 44$$

$$s.e.[median] = \frac{s.e.[\hat{S}]}{pdf[median]} = \frac{0.1813}{44} = 0.0041$$

where 0.1813 is the Greenwood estimate of the standard error at the observation selected to be the median.

Once the standard error is estimated, a z-interval is computed as the confidence interval for the median survival time (the median of flipped data). The 95% confidence interval using a t-statistic of 1.96 is computed as:

$$[0.016 - 1.96 \bullet 0.0041, \quad 0.016 + 1.96 \bullet 0.0041] = [0.008, 0.024]$$

Klein and Moeschberger (2003) discourage use of the Greenwood pdf-Z interval, calling equation 7.28 a "crude" estimate of the pdf, sufficiently unreliable for small sample sizes that they avoid its use and compute intervals by the inverted sign test method. This second type of interval inverts a sign test for multiply-censored data to avoid estimating the pdf, was originally developed by Brookmeyer and Crowley (1982), and so is called the B-C sign method.

The B-C sign method computes an estimate of the sign test statistic for multiply-censored data as a ratio whose variation is approximately normal (equation 7.30). This ratio at the center of equation 7.30 is computed for each detected observation in the K-M table. All observations whose ratios lie between the critical Z statistics at each end are considered inside the sign-test confidence interval. The extreme observations still within the limits of the Z statistics become the endpoints of the $(1-\alpha)\%$ interval. The B-C sign equation (Klein and Moeschberger, 2003) is

$$-z_{1-\alpha/2} \leq \frac{\hat{S} - (p)}{s.e.[\hat{S}]} \leq +z_{1-\alpha/2} \tag{7.30}$$

where $\hat{S}$ is the estimated survival probability for each detected observation, p is the percentage of the target percentile at the center of the interval, and s.e.$[\hat{S}]$ is Greenwood's standard error given in equation 6.6. $\hat{S}$ and s.e.$[\hat{S}]$ are printed for each observation by K-M software. It should be noted that the percentage p is in the same direction as the survival probabilities, and as are the original concentration data prior to flipping. So the 25th percentile of concentration is the observation with a 25 percent survival probability, and is the $(1-p) = 75$th percentile of the flipped Time variable. Table 7.1 lists the B-C sign test statistic for the median in the column "B-C Sign", calculated at each uncensored observation (here the flipped blood lead data) using equation 7.30.

For a two-sided 95% confidence interval, 2.5% of the error is placed on each side of the interval. The $(1-\alpha/2)$ Z (standard normal) statistic is Z(0.975) or 1.96. Any observation with a value in the B-C Sign column between and including $-1.96$ and $+1.96$ is within the 95% inverted sign test confidence interval for the median. Because the B-C Sign statistic jumps in value from one detected observation to the next, the confidence interval should also include part of the region extending to the first observation with a B-C Sign statistic greater in absolute value than 1.96 (unless the statistic for the endpoints were exactly equal to 1.96). To account for this, Brookmeyer and Crowley (1982) use the convention that the interval shall include the first observation on the high Time side with absolute value of its statistic > 1.96. This is the low side for concentration. From Table 7.2 the set of observations with B-C Sign statistics less than 1.96 in absolute value are the lead concentrations between 0.016 and 0.019 µg/l. Including the next observation on the high Time (low concentration) side, the lead concentration of 0.014 with B-C Sign statistic of –2.1, the 95% confidence interval on the median lead concentration is [0.014, 0.019] using the inverted sign test.

Klein and Moeschberger (2003) suggest transforming the sign test ratio when determining which observations are within the (1–α)% interval. Two alternative transformations are log-log and arcsine transformations (Borgan and Liestøl, 1990). The choice of these transformations arose out of the shape of hazard functions in survival analysis for producing confidence intervals on probabilities, rather than intervals for survival times. Borgan and Liestøl (1990) claim that more accurate coverage probabilities (1–α) are obtained using one of these two transformed test statistics for small data sets that follow a Weibull distribution. Their applicability has not yet been demonstrated for use in intervals for environmental data sets, whose shape is generally close to a lognormal distribution. So the original B-C sign test method should be used for environmental data until these variations have been tested further.

Table 7.3 summarizes the Greenwood pdf-Z and B-C sign test results for a 95% two-sided confidence interval around the median for the multiply-censored blood lead data. Also shown is the "binomial" confidence interval for singly-censored data, applied to the blood lead concentrations by treating all data below the highest detection limit of 0.02 as simply < 0.02. The bootstrap confidence interval is discussed in a later section. The B-C sign interval is shorter than the Greenwood pdf-Z interval, is fully nonparametric, and does not require a highly-variable estimate of the pdf of the distribution as does Greenwood. The usefulness and precision of the

**Table 7.2.** B-C inverted sign test statistics for determining confidence intervals for the median of the blood lead concentrations.

| Time | Lead | Survival Probability | Standard Error | B-C Sign |
|---|---|---|---|---|
| 0.731034 | 0.261 | 0.962963 | 0.036345 | 12.7380 |
| 0.822951 | 0.177 | 0.925926 | 0.050401 | 8.4507 |
| 0.825926 | 0.174 | 0.888889 | 0.060481 | 6.4299 |
| 0.893939 | 0.106 | 0.851852 | 0.068367 | 5.1465 |
| 0.950847 | 0.049 | 0.814815 | 0.074757 | 4.2112 |
| 0.966038 | 0.034 | 0.777778 | 0.080009 | 3.4718 |
| 0.975472 | 0.025 | 0.740741 | 0.084337 | 2.8545 |
| 0.976471 | 0.024 | 0.703704 | 0.087877 | 2.3180 |
| 0.981356 | 0.019 | 0.527778 | 0.166001 | 0.1673 |
| 0.984483 | 0.016 | 0.351852 | 0.181330 | -0.8170 |
| 0.985714 | 0.014 | 0.175926 | 0.153932 | -2.1053 |
| 0.986275 | 0.014 | 0.000000 | 0.000000 | * |

**Table 7.3.** Nonparametric two-sided 95% confidence intervals on the median blood lead concentrations. The binomial interval only handles one detection limit.

| Method | Lower Limit | Median | Upper Limit |
|---|---|---|---|
| Binomial (94.6%) | < 0.02 | < 0.02 | < 0.02 |
| Greenwood pdf-Z | 0.008 | 0.016 | 0.024 |
| B-C Sign | 0.014 | 0.016 | 0.019 |
| Bootstrap K-M | 0.014 | 0.016 | 0.019 |

binomial interval is greatly diminished by its requirement of increased censoring – the only information is that the entire interval is less than 0.02. Of the three, the B-C sign interval provides the greatest precision and information content for multiply-censored data.

*Nonparametric confidence intervals for percentiles other than the median*

Nonparametric interval estimates for percentiles other than the median can be computed by the same methods used for the median – intervals based on binomial probabilities, Greenwood pdf-Z, the B-C inverted Sign interval, and bootstrapping. With binomial intervals all data below the highest detection limit (HDL) are again treated as <HDL. The binomial table provides interval endpoints, where p is the percentage related to the percentile of interest at the center of the interval.

For other percentiles than the median the Greenwood pdf-Z equation 7.27 becomes equation 7.31

$$s.e.[p] = \frac{s.e.\left[\hat{S}_p\right]}{pdf[p]} \qquad (7.31)$$

where p is the survival probability or percentage for the percentile at the center of the interval. The pdf is estimated using probabilities slightly larger and smaller than p, analogous to equation 7.28. The variability in $\hat{S}_p$ is assumed to follow a normal distribution as in equation 7.29. This will be a poorer approximation at extreme percentiles near 0 and 1.

The B-C inverted sign test interval of equation 7.30 can be used without modification for confidence intervals on other percentiles. $\hat{S}$ is evaluated at the pth percentile location, not at the median. The percentage p is in the same scale as are survival probabilities and original concentrations, and will correspond to the (1−p)th percentile of the flipped Time variable.

These intervals are briefly illustrated by computing a one-sided upper 95% confidence bound on the 90th percentile of the lead data. This is a value in which there is 95% confidence of being exceeded in no more than 10% of the population.

Endpoint positions for the binomial interval are found by entering a binomial table with probability p = 0.9, along with the sample size and confidence coefficient.

For a 95% upper bound the input constant is the confidence coefficient, 0.95. The output from Minitab's® table is

```
Binomial with n = 27 and p = 0.9

 x    P( X <= x )      x    P( X <= x )
26      0.941850      27              1
```

There is a 94.18% probability of being less than or equal to the 26th observation, and so a (1−0.9418) = 6% chance of exceeding this value. If this is close enough to 5%, the 26th observation from the low end, or 0.177, is the endpoint for a 94% non-parametric upper confidence bound on the 90th percentile. As this lead concentration is well above the highest detection limit, no confusion is caused by the censoring, and more complex intervals are not necessary. But to illustrate the other intervals we go on.

The K-M estimate of the 90th percentile is the observation with the highest survival probability $\leq 0.9$. This is the observation with $\hat{S} = 0.888$, and Lead = 0.174. A 95% upper bound on this value using Greenwood's pdf-Z interval is

$$s.e.[p = 0.9] = \frac{s.e.[\hat{S}]}{pdf[0.9]} = \frac{0.0605}{0.6820} = 0.0887, \text{ where}$$

$$pdf[0.9] = \frac{\hat{S}(T^+(0.95)) - \hat{S}(T^-(0.85))}{T^-(0.85) - T^+(0.95)} = \frac{0.9630 - 0.8519}{0.8939 - 0.7310} = \frac{0.1111}{0.1629} = 0.6820$$

and so the 95% upper bound on the 90th percentile is

$$\hat{T}(0.9) + z_{1-\alpha} \bullet se[median] = 0.174 + 1.64 \bullet 0.0887 = 0.319$$

This bound is much higher than the binomial endpoint, lending support to caution when using this method for smaller sample sizes and percentiles close to the endpoints of the distribution.

The B-C Sign endpoint is computed using a one-sided version of equation 7.30 and p = 0.9 rather than 0.5 for the median.

$$\frac{\hat{S} - (0.9)}{s.e.[\hat{S}]} \leq +z_{1-\alpha}$$

From Table 7.4, comparing the B-C Sign statistic to the (1−α) normal quantile $Z_{0.95} = 1.64$, the one-sided 95% upper confidence bound on Lead is at the observation whose B-C Sign statistic is the largest value < 1.64, or the second highest observation of 0.177. This agrees with the binomial interval.

## Bootstrapped intervals

An alternative method of obtaining unbiased nonparametric estimates of parameters and their confidence intervals is called bootstrapping (Efron, 1981). Bootstrap estimates are produced by computing statistics on repeated random samples taken with replacement from the observed data. The repeated samples have the same number of observations as in the observed data set. Censored and uncensored observations are equally available for sampling. The summary statistic is computed for each random sample either by Kaplan-Meier or another censored data procedure. The distribution of the statistic is then estimated by the percentiles of the collection of computed values. For example, if 1000 means were computed by performing robust ROS 1000 times, a two-sided 95% confidence interval for the mean could be determined from the $2.5^{th}$ and $97.5^{th}$ percentiles of these means, the $25^{th}$ and $975^{th}$ of the 1000 mean values when ordered from low to high. Akritas (1986) and Efron (1981) demonstrated the utility of bootstrap estimates for censored data.

To continue our example of estimating the median blood lead level in herons (Golden et al., 2003), the median along with a 95% confidence interval is estimated without assuming a distributional shape for the data by bootstrapping. To begin with, the median of the original 27 observations can be calculated using the Kaplan-Meier (K-M) procedure in Minitab®. The output was listed in Table 7.2. The median is the observation whose survival probability is the largest value less than or equal to 0.5. For these 27 observations, that corresponds to a Lead concentration of 0.016 after rounding.

The same method can be repeatedly used on random samples from these sample data in order to provide a bootstrap estimate of the median, and a 95% nonparametric confidence interval around this estimate. Bootstrapping methods are beginning to be provided in statistical software, and if not present, are easy to add with a macro program. The Minitab® macro for bootstrapping Kaplan-Meier estimates is called KMBoot, and is invoked by the command

%KMBoot c1 c2

where c1 is the column of left-censored observations and c2 is the column of censoring indicators. One thousand random samples from the data in c1 are created by sampling with replacement, and temporarily stored within the macro. That is, a sample of 27 observations is selected from the original 27 data values. Each of the 27 values has an equal chance of being selected at every step, so some of them are chosen more than once for the new sample. For each of the 1000 new samples, an estimate of the median is computed using Kaplan-Meier. The median of these 1000 estimates is the bootstrap estimate of the K-M median, and as shown by the output below is approximately equal to 0.016, the original sample median. The 95% confidence interval for the bootstrap K-M estimate is formed from the $25^{th}$ and $975^{th}$ smallest estimates out of the 1000 estimates computed. The results of the KMBoot macro are printed below, and shown in Figure 7.6

```
Endpoints of 90%, 95%, 99% confidence intervals
           based on bootstrap samples
                of the K-M median

Bootstrap Kaplan-Meier median = 0.015517
```

**Table 7.4.** Calculations for a one-sided upper bound on the 90th percentile using the B-C method.

| Time | Lead | Survival Probability | Standard Error | B-C Sign |
|---|---|---|---|---|
| 0.731034 | 0.261 | 0.962963 | 0.036345 | 1.73237 |
| 0.822951 | 0.177 | 0.925926 | 0.050401 | 0.51438 |
| 0.825926 | 0.174 | 0.888889 | 0.060481 | -0.18370 |
| 0.893939 | 0.106 | 0.851852 | 0.068367 | -0.70425 |
| 0.950847 | 0.049 | 0.814815 | 0.074757 | -1.13949 |
| 0.966038 | 0.034 | 0.777778 | 0.080009 | -1.52759 |
| 0.975472 | 0.025 | 0.740741 | 0.084337 | -1.88836 |
| 0.976471 | 0.024 | 0.703704 | 0.087877 | -2.23375 |
| 0.981356 | 0.019 | 0.527778 | 0.166001 | -2.24228 |
| 0.984483 | 0.016 | 0.351852 | 0.181330 | -3.02292 |
| 0.985714 | 0.014 | 0.175926 | 0.153932 | -4.70384 |
| 0.986275 | 0.014 | 0.000000 | 0.000000 | * |

```
***************************
Bootstrap estimates of the 90% confidence interval of the median
   UPPER90 also = UCL95, the upper 95% CI on the median.

Row     LOWER90      UPPER90
 1    0.0142857    0.0186441
***************************
Bootstrap estimates of the 95% confidence interval of the median

Row     LOWER95      UPPER95
 1    0.0142857    0.0186441
***************************
Bootstrap estimates of the 99% confidence interval of the median

Row     LOWER99      UPPER99
 1    0.0137255    0.0235294
***************************
```

Note that because Kaplan-Meier selects an observed value as its estimate of the median rather than interpolating, the possible values for the median are few in number. This is characteristic of procedures that do not assume a distribution for the sample statistic, but instead use the observed data repeatedly. It is especially true for data sets with a relatively modest number of observations, as with the 27 observations available here. The 95% confidence interval on the median spans values between 0.014 and 0.019 µg/L. This agrees well with the B-C inverted sign test interval presented previously (Table 7.3). Bootstrapping is certainly expected to be a better procedure than either the Greenwood pdf-Z method, or the binomial procedure that requires censoring to the highest detection limit. Either the bootstrap or

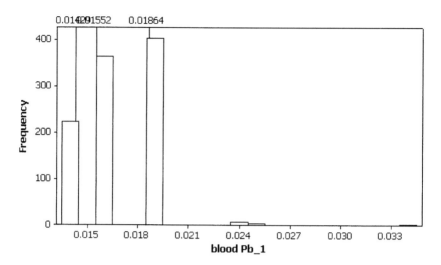

**Figure 7.6.** Histogram of 1000 bootstrap estimates of the median blood lead concentrations using Kaplan-Meier.

B-C inverted sign test methods can provide efficient nonparametric estimates for the median and other percentiles, along with their confidence intervals, for multiply-censored data.

*For further study*

Confidence intervals for Kaplan-Meier estimates are an active area of research. The probability intervals printed by Minitab's® output are "pointwise" intervals to be used only around a single probability estimate. When the entire pattern of the survival function is of interest, joint confidence bands analogous to multiple comparisons for ANOVA must be computed instead. More information on joint confidence bands is found in Chapter 4 of Klein and Moeschberger (2003), as well as in Nair (1984), Weston and Meeker (1991), and Jeng and Meeker (2001). Methods for computing intervals by inverting likelihood-ratio tests and other procedures have been developed by Emerson (1982), Simon and Lee (1982), Murphy (1995), and Slud and others (1984), among others. Akritas (1986) discusses bootstrapping to compute K-M confidence intervals in far greater detail than given here.

# Exercises

**7-1**    Using the zinc data from the Alluvial Fan zone of Millard and Deverel (1988) found in the data set CuZn, compute a 95% confidence interval on the mean zinc concentration, assuming the data follow a normal distribution. Based on a probability plot, is this assumption reasonable? If not, estimate a 95% confidence interval using bootstrapping and by using Land's method. How do these two intervals compare?

**7-2**    Estimate a 90% nonparametric confidence interval around the median of the zinc data using the binomial method (first re-censoring at the highest detection limit). Then compute the interval using the B-C sign method. How do these intervals compare? Which would you choose to use?

**7-3**    Construct a flow chart of the methods of this chapter for computing confidence intervals. Ignore methods known to be inadequate, such as the Greenwood pdf-Z method. Make sure that determining whether the data follow a normal or lognormal distribution figures prominently in your chart. Which method could appear throughout the chart, and work well regardless of the shape of the data distribution?

# 8

## WHAT CAN BE DONE WHEN ALL DATA ARE BELOW THE REPORTING LIMIT?

Measured concentrations may be consistently low in comparison to analytical detection limits, and all observations recorded as below the reporting limit. Some scientists have felt in these circumstances that they have no data to work with, because they have no single numbers reported to them. Is this correct? Absolutely not! For certain questions there is a lot of information in the fact that data are consistently below one or more thresholds. This information is contained in the proportion of data below versus above the specified thresholds. Methods based on binomial probabilities look at the probabilities of being above and below one or more thresholds. They can be used to answer several relevant questions about the data at hand.

For example, suppose two years of monthly samples of drinking water delivered by a supply system were tested for arsenic concentrations, and all 24 samples had concentrations reported as <3 μg/L. Three types of questions for which these data provide answers are

1.  Point estimates – What is an estimate of the typical arsenic concentrations being delivered in these waters?
2.  Exceedance probability of the reporting limit – How often might a detected concentration above 3 μg/L be measured in next year's samples?
3.  Exceedance probability for a standard higher than the reporting limit – How confident can we be that if current conditions are maintained, the drinking water standard of 10 μg/L will not be exceeded in more than 5 percent of samples?

Each of these three types of questions can be addressed when the data set consists of all nondetects.

With the exception of the example in the next section, there is no distinction made in this chapter between detection and reporting limits. The terms 'nondetects' and 'detects' refer to data below and above the reporting threshold. This is consistent with their (overly relaxed) use by most environmental scientists.

### Point estimates

There are at least two ways to compute an estimate of the mean or median for data sets that are entirely nondetects. A point estimate for the mean or median can (but should not) be computed using MLE when all data are below reporting thresholds. The MLE estimate will be extremely unreliable. As a second approach, a nonparametric estimate of the median can be calculated by stating that the median is below the reporting limit.

The (artificial) data set AsExample.xls consists of 13 measurements below the detection limit of 1, and so reported as within the interval of 0 to 1, and 11 measurements between the detection and quantitation limits and reported as within the interval of 1 to 3. Using maximum likelihood and assuming a lognormal distribution, summary statistics estimated for these data include:

| | Estimate | Standard Error | 95.0% Normal CI Lower | Upper |
|---|---|---|---|---|
| Mean (MTTF) | 0.994679 | 30255.3 | 0 | * |
| Standard Deviation | 0.121099 | 1839229 | 0 | * |
| Median | 0.987388 | 188559 | 0 | * |
| First Quartile (Q1) | 0.909821 | 1293760 | 0 | * |
| Third Quartile (Q3) | 1.07156 | 1114491 | 0 | * |
| Interquartile Range (IQR) | 0.161747 | 2408251 | 0 | * |

Summary statistics for the same data, assuming a normal distribution, are:

| | Estimate | Standard Error | 95.0% Normal CI Lower | Upper |
|---|---|---|---|---|
| Mean (MTTF) | 0.988121 | 25147.3 | -49286.8 | 49288.8 |
| Standard Deviation | 0.113527 | 240337 | 0 | * |
| Median | 0.988121 | 25147.3 | -49286.8 | 49288.8 |
| First Quartile (Q1) | 0.911549 | 187252 | -367006 | 367008 |
| Third Quartile (Q3) | 1.06468 | 136957 | -268430 | 268433 |
| Interquartile Range (IQR) | 0.153145 | 324209 | 0 | * |

The most important numbers to notice in these tables are the confidence intervals. When all observations are below the reporting limit the reliability of MLE parameter estimates is very poor. Confidence intervals are either undefined or extremely wide, showing that in reality there is insufficient information to compute reliable estimates of parameters such as the mean and standard deviation. Estimates will strongly depend on which distribution has been selected, even though observed data give no information on which distribution might be most appropriate. If an estimate known only as somewhere between –49000 and +49000 is considered too imprecise, individual parameter estimates by MLE for such data should be avoided.

The median of data that are all nondetects can always be stated as being <rl, where rl is the median reporting limit when ranked in order from low to high limits. A similar procedure is possible for other percentiles. For the AsExample data, the median of 24 observations would fall at the value midway between the 12th and 13th ranked observations. Since there are 13 values known to be <1, the median of these 24 observations is <1. No assumptions are made with this approach, and what is known about a single value is accurately presented.

In general, it is not very helpful to calculate point estimates for data that are entirely nondetects. Instead, inferences can be made concerning probabilities of exceeding the reporting limit(s), based on binomial probabilities. And if one group of data is entirely nondetects while another has some detects, tests of differences in the probabilities of exceedance can be performed.

## Probability of exceeding the reporting limit

When all data are nondetects the observed proportion of data exceeding the reporting limit is 0%. However, this is a sample percentage, only an estimate of the percentage above the reporting limit for the underlying population. If p is the proportion of the population below the reporting limit, 1–p is the proportion above. The certainty with which p is known is a function of the number of observations n. As n increases, the confidence interval around p decreases. The methods of this section involve estimating a confidence interval around the population proportion p, based on binomial probabilities.

Answers to the following four questions illustrate what can be accomplished when all data are nondetects. These questions come in pairs, with questions 1 and 3 estimating a confidence interval for p, and questions 2 and 4 testing to see if a specific value for p is within a confidence interval. The first two questions deal with the proportion p, while the second two questions deal with numbers of occurrences (p•n).

   1. What is the range of possible proportions of nondetects actually in the population from which these data came?

The sample estimate for the proportion of data below the detection limit is $p = c/n$, where c is the number of nondetects and n is the sample size. The estimated proportion above the detection limit is $1 - (c/n)$, When all measured values are below the detection limit, $c = n$ and the observed proportion below is $p = 1$. A two-sided confidence interval around p is constructed using quantiles of the F distribution (Hahn and Meeker, 1991).

$$[\text{LL, UL}] = [(1 + (n-c+1)\ F1\ /\ c)-1, \quad (1 + (n-c)\ /\ (c+1)\ F2)-1] \qquad (8.1)$$

where   LL   =   Lower limit of the confidence interval, = 0 when c = 0.
        UL   =   Upper limit of the confidence interval, = 1 when c = n.
        F1   =   $F_{(1 - \alpha/2,\ 2n - 2c + 2,\ 2c)}$ , and
        F2   =   $F_{(1 - \alpha/2,\ 2c + 2,\ 2n - 2c)}$

When all measurements are nondetects, so that c = n and the observed probability equals the maximum of 1, all error must be placed on one side of the interval, and $F_{1 - \alpha}$, rather than $F_{1 - \alpha/2}$ is used in the above equations.

For the example of c=24 nondetects in two years of monthly measurements of arsenic in drinking water (n=24), a two-sided 95% confidence interval for the proportion of arsenic concentrations below the reporting limit of 3 μg/L is:

[LL, UL]          $= [(1 + (24-24+1) F_1 / 24)^{-1}, (1 + (24-24) / (24+1) F_2)^{-1}]$
                  $= [(1 + F_{(0.95, \, 2, \, 48)} / 24)^{-1}, ( \, 1 \, )^{-1}]$
                  $= [(1 + 3.19 / 24)^{-1}, (1)]$
                  $= [0.88, 1]$

At a 95% confidence level, between 88 and 100 percent of arsenic concentrations in the population of delivered drinking waters can be expected to be below the detection limit of 3 μg/L, based on the data collected. No more than 12 percent of measurements are expected to exceed 3 μg/L.

In Minitab®, this confidence interval can be obtained with the command

### Stat > Basic Statistics > 1 proportion

entering n = 24 as the number of 'trials', and c = 24 as X, the number of 'events':

```
                                                      Exact
Sample    X    N    Sample p        95% CI           P-Value
  1      24   24    1.000000   (0.882654, 1.000000)   0.000
```

The 95% confidence interval states that based on the evidence in these data, the true proportion of nondetects lies (with 95% confidence) between 88 and 100%. The p-value reported by Minitab® is for a test of whether the proportion of nondetects is significantly different than 0.5. The small p-value shows that it is different, but this is not a question of particular interest here.

2. Are fewer than 10% (or some other proportion) of values in the population above the detection limit?

The test of whether (1-p), the proportion of detects, equals 10% or less is identical to a test for whether p, the proportion of nondetects, is 90% or greater. The motivation for doing so is often the presence of some legal standard or corporate guideline – 10% or more detects is considered too high for some reason, and if true a change would need to be implemented to lower the proportion. The null hypothesis is assumed to be true until proven otherwise, and could be set to assume there is a problem unless the data prove otherwise (it also could be set as the reverse). Assuming the proportion is high until proven otherwise, the null hypothesis would state that there are 10% or more detects in the underlying population, and therefore fewer than 90% of observations below 3 μg/L. The alternative hypothesis is the statement to be demonstrated – there are fewer than 10% detects, or greater than 90% of observations in the population are below 3 μg/L. This one-sided test is conducted by Minitab® with the same command as above, after setting a directional option for the alternative hypothesis in the Options dialog box. The percentile of nondetects p is set to 0.9, and the alternative is set to greater than 0.9. The data measured (c=24) were the number of nondetects, so the null and alternative hypotheses are stated in terms of the proportion of nondetects expected. The result is

```
Test of p = 0.9 vs p > 0.9
                                    95%
                                   Lower        Exact
Sample     X     N    Sample p     Bound       P-Value
  1        24    24   1.000000    0.882654      0.080
```

This tells us again that the lower bound of the 95% confidence interval for p is 88%, and a p-value for the test is 0.08. Because the proportion 0.9 was entered, this test determines whether the proportion of nondetects equals 0.9, just as stated in the first line of the output. Interpreting the p-value, 24 out of 24 nondetections would occur about 8 percent of the time when the true proportion of nondetects is 0.9. Eight percent is larger than the traditional 5% significance level, so the null hypothesis that there are 10% or more detections cannot be rejected. If we need to prove that there are fewer than 10% detections for legal or other purposes, more data are required. For example, after 6 more months and with all 30 observations below the detection limit, the result for the same test would be:

```
Test of p = 0.9 vs p > 0.9
                                    95%
                                   Lower        Exact
Sample     X     N    Sample p     Bound       P-Value
  1        30    30   1.000000    0.904966      0.042
```

and since $p < 0.05$ the null hypothesis is rejected. As stated by the alternative hypothesis, fewer than 10% detects are expected from this population. Also note the direct correspondence between the p-value and the confidence bound. When the 95% (0.95) lower confidence bound is below (and so includes) the tested proportion of 0.9 as a possible proportion, the p-value will be above $(1-0.95) = 0.05$. When as with the 30 observations the confidence bound is higher than (does not include) the tested proportion of 0.9, the p-value is below 0.05.

3. How many values above 3 µg/L can be expected in a new set of 12 observations from this same population?

To answer this question a confidence interval on the expected proportion of detections is computed, and the endpoints multiplied by the number of new observations to obtain the expected range in numbers of new detections. For the 24 arsenic nondetects the two-sided 95% confidence interval [LL, UL] for the proportion of nondetects p was between 88 and 100%. The expected proportion of detected values $(1-p)$ is therefore between 0 and 12%. For the $n_{NEW} = 12$ monthly samples of arsenic to be collected in the coming year, assuming the concentrations of arsenic are not changing, the number of expected 'events' (NEE) equals the number of new observations times the probability of that event occurring, or

$$NEE = n_{NEW} * (1-p)$$
$$= 12 * [0 \text{ to } 0.12] = [0 \text{ to } 1.4]$$

Between 0 and 1.4 of the 12 new samples can be expected to have a concentration above 3 μg/L.

4. What is the probability that more than 1 (or some other number of) detected value(s) occur in the next 12 (or some other number of) samples?

This question specifies the number of detects as a 'standard' or 'quality control measure'. One detect out of the next 12 will be allowed; more than 1 will not be. Of interest is whether the probability of failing the standard is less than a specified risk level $\alpha$. If based on the 24 existing observations the probability of failing the standard in the next 12 observations exceeds $\alpha$, some treatment might be implemented right away. Assume $\alpha$ is the traditional risk level of 5 percent. The probability of measuring 2 or more detects out of 12 new samples can be calculated with binomial tables.

To compute the probability of getting 2 or more detects, first compute the 95% confidence interval on the proportion of detects using the 24 existing values. From the answer to question 1, between 0 and 12 % of new data can be expected to be detects, with 95% probability. The two percentages 0 and 12, the endpoints of the 95% confidence interval on (1-p), will be used to compute the risk of failing the standard.

For the percentage (1–p) = 12% detects, the probability of measuring 2 or more detects out of the next 12 measurements is

$$\text{Prob } (x \leq 1) = \text{Binomial } (1, n_{NEW}, [1-p]) = \text{Binomial } (1, 12, 0.12)$$

where the function Binomial is the binomial cumulative distribution function, equation 8.2:

$$\text{Binomial } (x', n, p) = \text{Prob } (x \leq x') = \sum_{i=0}^{x'} \left( \frac{n!}{i!(n-i)!} \right) p^i (1-p)^{n-i} \qquad (8.2)$$

where $n! = n \cdot (n-1) \cdot .... \cdot 1$.

Both of the 95% interval endpoints, 0 and 0.12, are used as p in equation 8.2 to determine the range of probabilities. Therefore the probability that there will be 1 or fewer detected values in the next 12 samples is between

$$\sum_{i=0}^{1} \left( \frac{12!}{i!(12-i)!} \right) 0.12^i (0.88)^{n-i} = (1)1(0.88)^{12} + (12)0.12(0.88)^{11} = 0.569 \text{ for } p = 0.12$$

$$\sum_{i=0}^{1} \left( \frac{12!}{i!(12-i)!} \right) 0^i (1)^{n-i} = (1)1(1)^{12} + (12)0(1)^{11} = 1 \text{ for } p = 0$$

and so the probability of compliance, of getting 1 or fewer detects out of the 12 new observations, is between 57 and 100 %. There is somewhere between a 0 and 43% probability of non-compliance. This is understandable given that if the true proportion of detects is as high as 12% (the upper end of the confidence interval), the expected number of detects is 1.4 out of 12, and so it should not be very unusual to see 2 detects. But the 43% probability is higher than the traditional risk level of 5%.

More information on computing binomial confidence, tolerance and prediction intervals is found in Hahn and Meeker (1991).

### Exceedance probability for a standard higher than the reporting limit

Observing that all data fall below the reporting limit provides information about the likelihood of exceeding a legal limit at or higher than the reporting limit. If the value is below the reporting limit, it is also below the legal standard. Two approaches may be taken. In the first and simplest, the data are recorded as either exceeding or not exceeding the legal limit and the binomial methods of the previous section applied. The advantage of doing so is that no distribution need be assumed for the data. The disadvantage is that the proportion of data exceeding the reporting limit is only an upper bound on the proportion of data exceeding the standard. If there are no more than 12% detects, as in the previous section, then the probability of exceeding a higher legal limit is also no more than 12%. But it may be considerably smaller than 12%. How much smaller than 12% is unknown unless it is reasonable to assume a distribution for the data, allowing the difference in probabilities between exceeding the reporting limit and exceeding the legal limit to be modeled. This is the second approach.

*Binomial (nonparametric) tests*
The binomial test determines whether a percentile of the data distribution exceeds the legal limit at a stated confidence level. For the special case of testing the median or $50^{th}$ percentile the test is also called the sign test, and is discussed in detail in Chapter 9.

To test whether the median of 24 arsenic concentrations, all below 3 µg/L, exceeds the legal limit of 10 µg/L, the null hypothesis is stated as the proportion of data above the legal limit of 10 is 0.50. In other words, the median is at the legal limit. The binomial test uses the number of times ('events') the data exceed the limit out of the total number of 'trials'. If there were many more exceedances than 50%, the null hypothesis would be rejected. In the arsenic example, all of the 24 'trials' are below both the detection limit and the legal limit – there are 0 exceedances. Using the 1 Proportion routine within Minitab® and testing the proportion of detects = 0.5 versus the alternative that it is greater than 0.5, the test results are

| Sample | X | N | Sample p | 95% Lower Bound | Exact P-Value |
|--------|---|----|----------|-------|---------|
| 1 | 0 | 24 | 0.000000 | * | 1.000 |

With a p-value of essentially 1, the binomial test states that there is no reason to reject the null hypothesis. There is no evidence that the median concentration is above the legal limit of 10. With all the observations below the detection limit, this is not a surprising result.

If instead of the median a regulation states that another percentile of the distribution shall not exceed the standard, the quantile test (Conover, 1999) may be used. The quantile test is the binomial test applied to a proportion other than 50%, where the quantile = the percentile /100. So the 50th percentile is the 0.5 quantile. Suppose that a regulation asserts that 90% or more of observations must be below the limit – the proportion of exceedances must be no more than 10%. The null hypothesis (of compliance) states that there are 10% or fewer exceedances, p = 0.1, with the alternative hypothesis that there are greater than 10% exceedances. The binomial test is run again using the proportion tested equal to 0.1 (10% exceedances) and the 'event' being an exceedance, of which there are none. The result is

```
Test of p = 0.1 vs p > 0.1
                                 95%
                               Lower      Exact
Sample    X    N    Sample p   Bound      P-Value
  1       0   24    0.000000     *        1.000
```

and so there is also insufficient evidence to reject the null hypothesis and declare that the legal standard of no more than 10% detects has been violated. This same test can be conducted using the qtl macro, whose output is perhaps more clear. The qtl macro performs the quantile test, and adds text interpreting the result. The macro is invoked by

%qtl c2 90 10

where the first argument (here = c2) is the column of data, the second argument (here = 90) is the percentile to be tested, and the third argument (here = 10) is the value for the legal standard. The output of the macro is:

```
Is the probability of exceeding 10 greater than 10 percent?

Test and CI for One Proportion
Test of p = 0.1 vs p > 0.1
                                 95%
                               Lower      Exact
Sample    X    N    Sample p   Bound      P-Value
  1       0   24    0.000000     *        1.000

There were 0 out of 24 observations greater than 10.
To reject p = 0.1 with no more than 5% error, more
than 5 of 24 observations must be greater than 10.
```

If the true proportion were 10% exceedances, 2.4 exceedances would be expected. The macro states that 5 exceedances would need to be observed to have enough evidence to prove that p was at least 10%, with 24 observations.

Of course the test against a standard could also be run assuming non-compliance, by reversing the direction of the alternate hypothesis. The burden of proof would be to determine if 0 out of 24 exceedances were enough evidence to declare that the probability of exceedance is less than 10%. The Minitab® output for this perspective is:

```
Test and CI for One Proportion
Test of p = 0.1 vs p < 0.1
                                    95%
                                  Upper        Exact
Sample    X     N    Sample p     Bound       P-Value
  1       0    24    0.000000    0.117346      0.080
```

The p-value for the test is 0.08. This indicates that the probability of observing 0 out of 24 exceedances just due to chance is 8% when the true proportion of exceedances is 10%. If the acceptable error rate $\alpha$ is 0.05, the null hypothesis of non-compliance is not rejected. The small p-value would indicate a preference towards compliance, but the strength of the evidence is insufficient. If there were 6 more samples all of which were nondetects, the test becomes:

```
Test and CI for One Proportion
Test of p = 0.1 vs p < 0.1
                                    95%
                                  Upper        Exact
Sample    X     N    Sample p     Bound       P-Value
  1       0    30    0.000000    0.095034      0.042
```

and non-compliance is rejected at the acceptable error rate $\alpha$ of 0.05.

*Parametric tests of exceedance*

The parametric approach to comparing all nondetects to a standard is to estimate the exceedance probability of a standard as a function of the difference between the detection limit and the legal standard. This requires an assumption about the shape of the data distribution. Smith and Burns (1998) present a method to estimate exceedance probabilities of a legal standard assuming a normal distribution when all data are nondetects. The assumption of a normal distribution is not commonly adhered to by environmental data. However, they state that it is more applicable for composite samples, where observations measured in the laboratory are the mean values of several individual samples composited together prior to analysis. Typically, 10 to 20 individual samples are combined into one composite sample in environmental studies. For the large skewness found in environmental data, composites of a small number of individual samples can still exhibit considerable skewness, so the assumption of a normal distribution for composite samples should be demonstrated.

A more reasonable assumption might be that data follow a lognormal distribution. In a brief side comment, Smith and Burns (1998) suggest a method for estimating exceedance probabilities when assuming data follow a lognormal distribution. The estimated lower bound for the percentile associated with the legal standard is a function of the difference between the standard (abbreviated Std) and the detection limit (DL), divided by the standard deviation $\sigma_L$, all in log units. This standardized distance is subtracted from the normal quantile of the binomial exceedance probability, the probability calculated in question 1 for exceeding the detection limit. Their estimator is

$$\Phi\left[\Phi^{-1}\left(\alpha^{1/n}\right) + \left(\ln[Std] - \ln[DL]\right)\Big/{\sigma_L}\right]$$
(8.3)

where $\Phi$ is the cumulative distribution function for the standard normal distribution and $(1-\alpha)$ is the desired confidence level for the lower bound. This equation assumes that data follow a lognormal distribution, but more importantly that a reasonable estimate of $\sigma_L$ can be obtained. When all data are below the detection limit, obtaining a reasonable estimate of $\sigma_L$ is unlikely. The Smith and Burns estimator in equation 8.3 comes from an un-refereed proceedings document rather than from a refereed journal article, and so should be approached with caution until validated by further work. Equation 8.3 agrees with their proceedings document. A minus sign rather than a plus sign appears between the two main components in Smith and Burns (1998); the minus sign is an error (D. E. Smith, personal communication, 2004).

Note that if a reasonable estimate of the standard deviation is doubtful, the first part of equation 8.3 is the binomial estimate of the lower confidence bound for the proportion of the population less than the detection limit, previously produced by the 1 proportion software:

$$\Phi\left[\Phi^{-1}\left(\alpha^{1/n}\right)\right] = \alpha^{1/n} = 0.05^{1/24} = 0.8826$$

This estimate requires no distributional assumption and so can be used instead of equation 8.3 as a lower bound for the proportion of the population less than the legal standard.

### Hypothesis tests between groups

The subject of hypothesis tests for differences in mean or median between two groups is covered at length in Chapter 9, and for three or more groups in Chapter 10. When one of the groups contains only nondetects, it can be a strong indication that there are differences among the groups. If so, the methods in those chapters can make that determination. If all groups consist entirely of nondetects, it is evidence that the distributions of data must be considered similar, at least within the analytical precision available to the scientist. With many nondetects it is difficult to determine whether data follow any specific distribution, so nonparametric tests are useful. Contingency tables (see Chapter 10) test differences between the proportions of detected observations among groups. These tests work well for one detection limit. The score tests of chapters 9 and 10 determine differences in the distribution functions of data with multiple detection limits, even if one group contains all nondetects. For testing among groups there is no need to avoid using a nonparametric test because of "too many" nondetects. If the proportion of nondetects differs significantly between groups, nonparametric score and contingency table tests will respond to that difference.

## Summary

When all observations are recorded below the detection limit, methods based on binomial probabilities can produce confidence intervals and hypothesis tests concerning the proportion and number of detects or nondetects in the population being measured. Though estimates of mean or median are not available, statements about the probability of exceeding the reporting limit or other threshold can be made. The quantile test can determine whether a specified percentile is proven to be above or below a legal standard, even when all observations are nondetects. When testing data of all nondetects against other data containing some detected observations, nonparametric methods such as contingency tables, the sign test, or the Kruskal-Wallis test (see Chapters 9 and 10) provide considerable power for determining whether one group generally produces higher values than another.

# Exercises

**8-1**    Thurman et al. (2002) measured concentrations of antibiotics in discharges from fish hatcheries across the United States. A summary of the data are found in hatchery.xls. Twenty-five samples contained no concentrations of tetracycline above the detection limit of 0.05 μg/L. Two samples did contain detectable concentrations, but these were believed to be analytical artifacts from another compound, and the observations were discounted. Based on twenty-five nondetects, and assuming that these hatcheries represent the conditions found at others to be sampled in the future, what is the likelihood of getting at least one detection in the next 15 samples analyzed?

**8-2**    Assuming these 25 locations reasonably represent fish hatcheries across the US, estimate a 90% confidence interval on the proportion of concentrations of tetracycline below 0.05 μg/L in waters draining fish hatcheries in the United States.

**8-3**    Use a contingency table analysis (see Chapter 10) to determine if the proportion of detections for oxytetracycline is significantly different than that for tetracycline.

**8-4**    MTBE in groundwater is a concern for drinking water supplies in states where the compound has been used as a gasoline additiive. If in a survey of a county's drinking-water supply wells, all 36 measurements have been recorded as below the detection limit of 3 ppb, the data are assumed to be lognormal, and the standard deviation of the logarithms (based on other data) is estimated to be 1.0, what is an estimated probability of exceeding the "level of concern" of 13 ppb in groundwater?

# 9

# COMPARING TWO GROUPS

In their classic paper on methods for censored data, Millard and Deverel (1988) studied copper and zinc concentrations in shallow ground-waters from two geologic zones underneath the San Joaquin Valley of California. Throughout this chapter these data will be used to determine whether zinc concentrations differ between the two zones. Several methods will be applied to the zinc data and their performance evaluated. Zinc concentrations were subject to two detection limits at 3 and 10 µg/L, with a number of values detected between these limits during the time when the lower detection limit was in effect. Using nonparametric methods discussed later in this chapter, Millard and Deverel found that zinc concentrations were different in the two zones. Methods that work reasonably well should also find this difference. Censored boxplots of the two data sets are shown in Figure 9.1.

Comparing two groups is a basic design in many environmental studies. In some cases, a 'treatment' group is compared to a 'control'. The control group represents background, or historical conditions. The treatment group represents conditions where, for example, contaminant concentrations are suspected to be higher, or numbers of healthy organisms lower. Differences are tested in one direction – treatment conditions are suspected to be worse in comparison to the control. Because a difference is expected in only one direction, these are called one-sided tests, tests where the direction of difference is specified as part of the study design. One-sided tests are also appropriate where the expected direction is an improvement over existing conditions – a new lab method with more accuracy, air concentrations following implementation of new scrubber technology, and so forth. The key to a one-sided test is not in which direction is expected, but that there is only one direction expected.

In other cases, the two groups are inspected for differences where either may be better or worse than the other. Neither group can be labeled as a control group. The interest is truly in whether measurement levels in the two groups are the same, or different. These are two-sided tests – differences are investigated in two directions. Two-sided tests are appropriate for comparing the concentrations in different locations, for example, where if either location has significantly higher values the outcome is of interest. The zinc data are like this – the question stated was whether the two groups had different concentrations of zinc – no direction was specified. For either one to be higher is of interest. This is a two-sided test.

For both one and two-sided tests, the procedures used when there is no censoring are familiar to data analysts – the parametric two-sample t-test, and the nonparametric Mann-Whitney (or its alternate name, the Wilcoxon rank-sum) test. When censored values are present, options for analysis include substitution methods, distributional methods, and nonparametric methods.

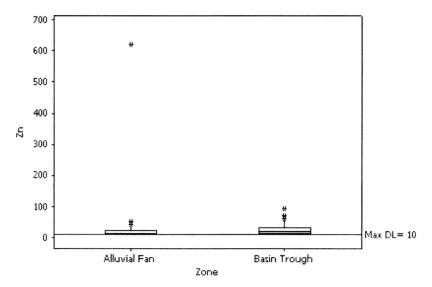

**Figure 9.1.** Boxplots of zinc (Zn) concentrations. Note that a greater percentage of the Basin Trough data are above the 10 ug/L limit.

## Substitution methods

The only way that a standard two-sample t-test can be run on data with nondetects is to fabricate (substitute) values prior to computing the test. One approach used in some environmental reports, but NOT recommended here, is to run the test twice, first substituting zero, and then the detection limit. The argument goes that if the results of the two tests agree, then perhaps these are the correct results. Perhaps, but this is far from a sure thing. Though the mean will change monotonically as the substituted value goes from the detection limit down to zero, the standard deviation may not. The t-test statistic is a function of both the mean and standard deviation, and will vary in an indeterminate pattern when different values are substituted. Regardless of which value is substituted, the process implies that more 'information' (the substituted values) is known about the data than is truly known by the analyst. If this 'information' is not correct, the test result will likely not be correct either.

Setting all nondetects equal to zero, the t-test does not indicate a difference between zinc concentrations within the two zones ($p = 0.995$):

```
Two-sample T for Zn0

Zone              N      Mean      StDev     SE Mean
Alluvial Fan     67      21.2      75.0       9.2
Basin Trough     50      21.3      19.3       2.7
```

```
Difference = mu (Alluvial Fan) - mu (Basin Trough)
Estimate for difference: -0.06
95% CI for difference: (-19.08, 18.97)
T-Test of difference = 0 (vs. not =): T-Value = -0.01
P-Value = 0.995  DF = 77
```

Setting all nondetects equal to their detection limits (some 3, some 10 µg/L), the t-test again indicates no difference between the groups (p = 0.869):

```
Two-sample T for Zn_dl
```

| Zone | N | Mean | StDev | SE Mean |
|------|-----|------|-------|---------|
| Alluvial Fan | 67 | 23.5 | 74.4 | 9.1 |
| Basin Trough | 50 | 21.9 | 18.7 | 2.6 |

```
Difference = mu (Alluvial Fan) - mu (Basin Trough)
Estimate for difference: 1.57
95% CI for difference: (-17.29, 20.43)
T-Test of difference = 0 (vs not =): T-Value = 0.17
P-Value = 0.869  DF = 76
```

Does this mean that there is no true difference between the two groups of zinc concentrations? No, it certainly does not. It was previously mentioned that the correct result is that the distribution of concentrations does differ between the two groups. Neither extreme of substituted value is likely to be close to the true concentrations actually present in the samples. As a third attempt at substitution, the common method of setting all nondetects equal to one-half of their detection limits is tried. This time, logarithms are taken following substitution to address the skewness of the data. Censored data are usually skewed, lowering the power of parametric t-tests to detect differences. A t-test on logarithms following one-half dl substitution again produces a non-significant p-value. The t-test fails to finds differences in the mean of the logarithms following substitution (p = 0.109):

```
Two-sample T for lnZn1/2
```

| Zone | N | Mean | StDev | SE Mean |
|------|-----|-------|-------|---------|
| Alluvial Fan | 67 | 2.444 | 0.816 | 0.10 |
| Basin Trough | 50 | 2.707 | 0.911 | 0.13 |

```
Difference = mu (Alluvial Fan) - mu (Basin Trough)
Estimate for difference: -0.263101
95% CI for difference: (-0.586284, 0.060081)
T-Test of difference = 0 (vs not =):T-Value = -1.62
P-Value = 0.109  DF = 98
```

Clearly, if the true result is that there are differences between the zinc concentrations in these two geologic zones, substitution followed by a t-test is an inadequate procedure for detecting it, regardless of the value being substituted.

### Maximum likelihood estimation

Standard parametric tests for differences between groups of uncensored data, such as Analysis of Variance (ANOVA) and t-tests, can be computed using simple linear regression where the explanatory variables are coded to indicate group membership. For the simple case of testing the difference between the mean of two groups, only one explanatory variable (X) is needed. X is a binary variable, coded as a 0 if the data come from the first group, and a 1 if the data are from the second group. Solving the regression produces an estimate for the slope of the X variable. This slope equals the difference between the two group means. The regression t-test of significance for this slope (assuming equal group variances) is the test to determine whether this difference equals zero. In this way, a two-sample t-test can be computed using regression software. This procedure can be used in computing a t-test for censored data.

Computation of parametric hypothesis tests between group means of censored data is accomplished using software for censored regression. Censored regression methods use maximum likelihood estimation to compute estimates of slope and intercept, and to conduct hypotheses tests on the significance of the slope coefficient. The benefits in using MLE methods include the fact that they work for data with multiple detection limits, and do not require substitution of fabricated data in order to perform the tests. The caution with MLE methods is that the validity of their results depends on selecting the correct distribution. For a small amount of censoring, the fit of data to the distribution can be evaluated with probability plots or hypothesis tests. For larger amounts of censoring, it is difficult to judge which type of distribution the data might have come from. In this case, use of parametric MLE methods requires that the choice of distribution be based on other knowledge, such as distributions for similar data in previous studies. For environmental data, the lognormal distribution is most often assumed when evidence is not available in the data themselves.

As noted in Chapter 6, the usefulness of MLE methods is limited if the sample size is small. In addition to the difficultly in determining whether small data sets follow a particular distribution, the optimization procedures of MLE do not have enough information with small data sets to accurately estimate parameters. As shown by many simulation studies (including Helsel and Cohn, 1988; Gleit, 1985) errors associated with MLE methods increase dramatically as sample sizes decrease below 30 or so observations. In this range, methods other than MLE are recommended.

To estimate a slope and intercept using maximum likelihood, possible values for the two parameters are adjusted until the values most likely to produce the observed measurements are determined. Matching of possible parameters to observations is done through the likelihood function. Parameters chosen by MLE maximize the likelihood function for censored regression, evaluated by setting the derivatives of the logarithms of the likelihood function with respect to each parameter equal to zero, simultaneously solving the two equations:

$$\frac{d \log L(\beta)}{d\beta_0} = 0 \qquad\qquad \frac{d \log L(\beta)}{d\beta_1} = 0 \qquad\qquad (9.1)$$

where $\beta_0$ is the intercept and $\beta_1$ is the slope.

MLE also estimates the standard errors of both parameters, so that confidence intervals around each can be constructed. Standard errors of coefficients are the square root of the entries on the main diagonal of the covariance matrix (Allison, 1995, p. 84). The distributional assumption for the procedure has a strong influence on the resulting confidence intervals, much as the normality assumption does for intervals around the parameters for simple least-squares regression. If a normal distribution is assumed, symmetry of the input data is critical for the results to make sense. If the input data are skewed, variance estimates will be too large, tests for parameters will be too often insignificant, and confidence intervals may have negative lower bounds when this is physically impossible. When these occur, a log or other transformation should be considered before using MLE.

Due to the skewness of many environmental variables, the assumption of normality for data is often not a good one. In many cases, the logarithms of data more closely follow a symmetric distribution than do the original data. In hypothesis testing, taking logarithms prior to conducting the test is a common practice, for uncensored as well as censored data. Yet doing so changes the hypothesis being tested. For two group tests in original units, the null hypothesis is that means of two distributions are identical. The alternative hypothesis is that they differ by an additive constant. With logarithms, the null hypothesis becomes a statement concerning the means of the logarithms of the data. The alternative hypothesis, that the mean logarithms are offset by an additive constant, translates to multiplication in the original units. When logarithms are used in a two-group test, the null hypothesis is therefore that the ratio of the geometric means of the original data equals one. The alternative is that this ratio is not equal to one. Finally, if the distribution of the logarithms is symmetric, as would be hoped in order to assume normality, the geometric mean estimates the median of the original units, not the mean. A two-group parametric test on the logarithms becomes a test for equality of medians. This is true whether the test is a standard t-test for uncensored data, or a censored regression with parameters computed by maximum likelihood.

*Example – the Zn data*

Censored regression is found in the survival analysis section of statistical software. Correct results for left-censored data require their entry as arbitrarily- or interval-censored values using the interval endpoints format. The endpoints span the range of possible values for censored data from 0 to the detection limit. In Minitab® this is accomplished by entering two columns as response variables. The explanatory variable is a column named GeoZone with 0 for data from the Basin Trough and 1 for data from the Alluvial Fan. The censored regression procedure, found in Minitab® under the Reliability/Survival menu, tests whether the mean zinc concentration in groundwater of the two zones differs, using maximum likelihood.

The output from the procedure is:

```
Estimation Method: Maximum Likelihood
Distribution: Normal
Relationship with accelerating variable(s): Linear

Regression Table
                       Standard                      95.0% Normal CI
Predictor   Coef        Error     Z       P       Lower      Upper
Intercept  21.6125     8.11997   2.66   0.008     5.69774   37.5274
GeoZone     0.762958  10.7311    0.07   0.943   -20.2697    21.7956
Scale      57.4123     3.75329                   50.5078    65.2607

Log-Likelihood = -596.254
```

The slope coefficient value of 0.76 estimates the difference between the two group means, and the Z statistic (0.07) tests whether 0.76 is significantly different from zero. The test is equivalent in function to the t-test for explanatory variables in simple least-squares regression. The p-value of 0.943 indicates that the null hypothesis of no difference cannot be rejected – there is insufficient evidence of a true difference in mean Zn concentrations between the two zones using this test.

This test assumed that both groups of Zn concentrations follow a normal distribution. That assumption can be checked using a probability plot, after the group mean is subtracted from both group's data. Differences between the observed data and their group mean are called residuals, and it is these residuals that are assumed to follow a specific distribution by parametric tests. The probability plot of residuals in Figure 9.2 appears nonlinear, indicating that the data do not follow a normal distribution. Therefore, the test might have failed to indicate a true difference because it had low power resulting from non-normal data.

Note that for software that does not include an interval-endpoint format for use with left-censored data, the data could be flipped and run as right-censored values. However, this will not produce the appropriate test. Right-censored survival analysis methods assume data have no upper bound. Values up to positive infinity convert back to concentrations down to negative infinity on the original scale; the lower bound of zero is not recognized. Without this lower bound, test results are incorrect. When using MLE procedures assuming a normal distribution, left-censored data must be entered as interval-censored values. However, as shown later, this is not the case for tests assuming a lognormal distribution.

Given that these data are right skewed, a log transformation will be employed prior to running another two-group test. The test will determine whether the mean of logarithms, (the geometric mean) differs between the two groups. Minitab® and other statistical software can do this step automatically by creating a likelihood function for the lognormal distribution. The assumed distribution in the dialog box is changed from Normal to Lognormal. For software that is unable to perform maximum likelihood for a lognormal distribution, data can be log-transformed and the normal distribution assumed for the logarithms.

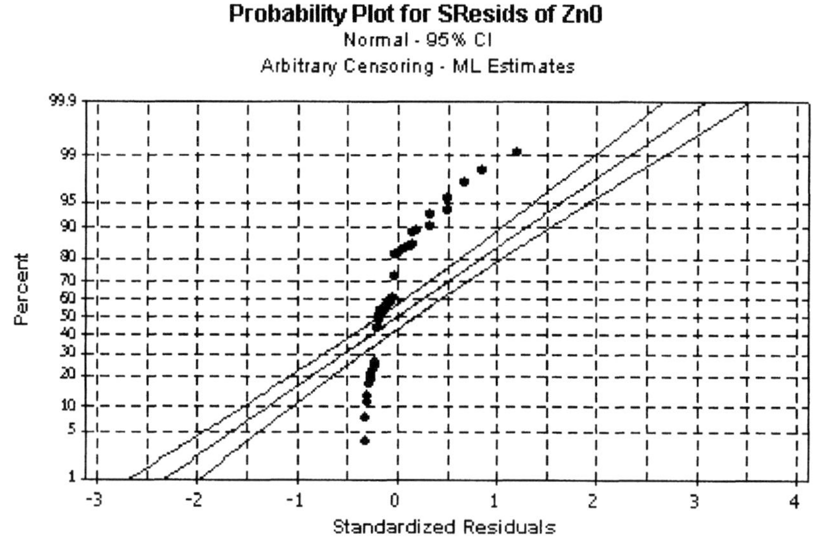

**Figure 9.2.** Probability plot of residuals from the censored regression of Zn concentrations against group membership. The data do not appear to follow a normal distribution.

For the zinc example the lognormal distribution is selected as the assumed distribution, and the results for the censored regression procedure are:

```
Estimation Method: Maximum Likelihood
Distribution: Lognormal
Relationship with accelerating variable(s): Linear

Regression Table
                    Standard                    95.0% Normal CI
Predictor    Coef      Error      Z      P       Lower      Upper
Intercept 2.72375   0.120325  22.64  0.000    2.48792    2.95958
GeoZone  -0.257408  0.161211  -1.60  0.110   -0.573377   0.0585604
Scale     0.842529  0.0618053                 0.729698    0.972806

Log-Likelihood = -407.296
```

The Z statistic for GeoZone has a much lower p-value (0.11) than when the normal distribution was assumed, but this value still leads to a conclusion of no difference between the two groups' geometric means. A probability plot (Figure 9.3) indicates that most data follow the straight line representing a lognormal distribution, though one or two points are outside the 95% confidence interval boundaries for the location of the distribution. Concern that there may be an effect due to violating the distributional assumption could lead to either use of yet another distribution, or to using a nonparametric approach.

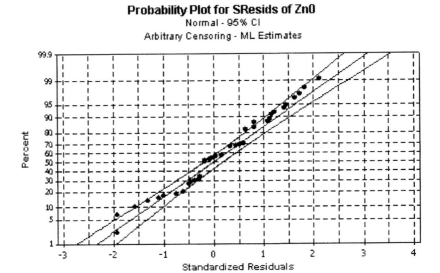

**Figure 9.3.** Probability plot for residuals from a regression of the Zn data versus group membership, compared to a lognormal distribution (center straight line).

When software does not allow input of interval-censored data, log-transformed data can be flipped and the procedure run. The result will be identical to a test on log-transformed data (or when assuming a lognormal distribution) had the interval endpoints format been available. The procedure is to take logarithms, flip the data by subtracting from a value larger than the largest logarithm, and run the right-censored regression. The lack of an upper boundary for right-censored data will map into negative infinity when the data are re-flipped back to the original log scale. Negative infinity for logarithms becomes a lower limit of zero when data are exponentiated back into units of concentration. For data with a lower bound of zero, censored regression can be correctly performed by software that does not allow interval-censored input by first taking logarithms, and then flipping to produce the required right censoring.

To illustrate the process when regression on right-censored data is the only option, the natural logs of Zn are computed, and then subtracted from 7, a number larger than the maximum logarithm, to produce right-censored data. The values are stored in a column named 'FliplnZn'. The indicator column format is used by the software to designate which values of FliplnZn are censored. The MLE regression output for the flipped natural logs of Zn includes:

```
Distribution: Normal
Response Variable: FliplnZn
```

```
Regression Table
                        Standard                  95.0% Normal CI
Predictor    Coef       Error       Z       P      Lower    Upper
Intercept    4.2763     0.1204     35.52   0.000   4.0403   4.5122
GeoZone      0.2575     0.1613      1.60   0.110  -0.0587   0.5736
Scale        0.84292    0.06188                    0.72996  0.97335
```

The slope coefficient for GeoZone and test results agree with those above when untransformed data were input as interval endpoints and a lognormal distribution was assumed.

In summary, for data with one or multiple censoring thresholds a two-group parametric test can be computed using censored regression, where a binary 0/1 variable is the only explanatory variable. Care must be taken, however, to judge the validity of the test's assumptions of a normal distribution and equal variance. Censored data are frequently non-normal. Violating the normality assumption will result in a loss of power to detect differences between the two groups. Environmental data are often fit better by lognormal distributions, and taking logarithms has the added benefit of allowing software with input only of right-censored data to correctly test data with a zero lower bound. However, the user must be aware that transforming data to logarithms changes the basic form of the test, determining whether the two groups' geometric means (medians) differ from a ratio of one.

## Nonparametric methods

Nonparametric methods do not require an assumption that data follow a specific distribution. They use no information on the shape of the distribution in conducting tests. Instead, data are ranked, providing information on the relative positions of each observation. Tests determine whether one group generally has more frequent high or low values, a test for whether percentiles of the data differ between the groups. For censored data, positions are represented by scores, which are ranks of the data adjusted for the information missing because some values are censored. Nonparametric tests using these methods are called score tests.

### One detection limit

The simplest situation for censored data is when there is only one detection limit. In this case, standard nonparametric tests such as the rank-sum test can be computed directly from the data. The position of each observation is represented by its rank. All nondetects are assigned the same value below the detection limit, a value less than the lowest detected observation. When the test converts data to their ranks, nondetects are represented as a tied rank lower than the rank for the lowest detected observation. These ranks will efficiently capture the information in the data, including the proportion of nondetects, accurately representing what is actually known about the data. Test results are reliable, not based on 'information' that is not known, and not dependant on the substitution of arbitrary values.

Evaluations of methods for testing differences between groups of censored data are scarce. However, in a noteworthy paper, Clarke (1998) evaluated 10 methods for comparing two groups of singly-censored data. Sample sizes were small, less than 10 observations in each group. Methods included substitution, and several "robust" methods estimating single values for nondetects by MLE or ROS (presumably using normal scores, but details were not included). Following the creation of artificial values for each censored observation, the two groups of data were evaluated by the Least Significant Difference (LSD) multiple comparison test, essentially a two-group t-test. In addition to the standard normal-theory test, the LSD test was computed on the logarithms of the data, and on the 'rankits', a rank transformation. The best test in the greatest number of cases was one using rankits following substitution.

Given the small sample sizes of the study, MLE methods may not have been expected to work well. But the good performance of substitution of a single number over "robust" methods seems to contradict the findings of Chapter 6. Are they contradictory? If so, why?

First, Clarke's paper does not make clear that simple substitution of a single value for nondetects followed by converting to rankits prior to testing with LSD results in a close approximation to the Mann-Whitney rank-sum test. Parametric tests on ranked data approximate their equivalent nonparametric procedures (Conover and Iman, 1981). Her best method was essentially a rank-sum test, which accurately represented what was known about the data – the nondetects were tied with each other and lower than detected observations. Second, all other methods somehow arbitrarily assigned values to censored data that were known only to be equal. In essence, Clarke found that any method of assigning unequal values to specific samples could produce a signal that wasn't in the original data. It could also obscure a signal that was present. Her paper should be strong evidence for using nonparametric methods on small censored data sets, and for using methods that do not estimate values for individual nondetected observations, when performing hypothesis tests.

*Rank-sum test for data with one detection limit*

Dissolved organic carbon (DOC) concentrations were measured by Junk et. al. (1980) in background wells and in other wells affected by cropland irrigation, in September of 1978 (Figure 9.4). Of interest is whether concentrations are higher in the wells affected by irrigation. Three values are censored at the single detection limit of 0.2 µg/L. The data are found in doc.xls and shown in Table 9.1.

The three nondetects would have had the ranks of 1 through 3 if analytical precision had allowed their concentrations to be quantified. However, with the available precision the three are tied with each other. Therefore each is assigned the rank of 2, the median of ranks 1 through 3. The Mann-Whitney (or rank-sum) test can be easily applied to these data, without any changes. This should be set up as a one-sided test – the question was "Are concentrations higher in wells affected by irrigation?" – so a difference in only one direction is of interest. The results below show

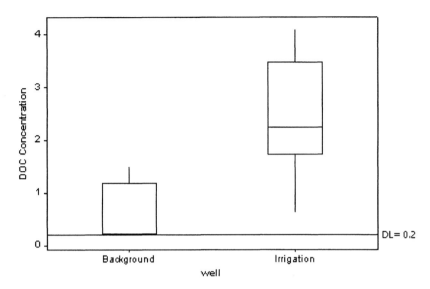

**Figure 9.4.** DOC concentrations in two types of wells (data from Junk et. al. 1980)

**Table 9.1.** DOC concentrations in two sets of wells. From Junk et. al. (1980).

| Background | Irrigation | Background Rank | Irrigation Rank |
|---|---|---|---|
| < 0.2 | 3.4 | 2 | 12 |
| 1.5 | 1.9 | 6 | 7.5 |
| < 0.2 | 3.7 | 2 | 13 |
| < 0.2 | 2.1 | 2 | 9 |
| | 3.2 | | 11 |
| | 2.4 | | 10 |
| | 1.2 | | 5 |
| | 4.1 | | 1 |
| | 1.9 | | 7.5 |
| | 0.6 | | 4 |

that DOC concentrations are higher in irrigation-influenced wells, with a p-value of 0.0064. No substitutions or assumptions of a distributional shape for these data were required to run the test.

```
Mann-Whitney Test and CI: Irrigation, Background

Irrigation   N = 10      Median =      2.250
Background   N =  4      Median =   <  0.200
Point estimate for ETA1-ETA2 is        2.000
96.0 Percent CI for ETA1-ETA2 is (0.601,3.700)    W = 93.0
Test of ETA1 = ETA2 vs ETA1 > ETA2 is significant at 0.0067
The test is significant at 0.0064 (adjusted for ties)
```

*Multiple detection limits.  Method I – censor at the highest limit*

With multiple detection limits, the rank-sum (Mann-Whitney) test can be run after censoring all values below the highest detection limit, representing them all as tied ranks. This is not the optimal method for censored data – information is lost when setting detection limits to a higher value, and when censoring detected observations measured between detection limits. However, it is the easiest method when advanced statistical software is not available, and may be sufficient to detect differences between groups. For the zinc data from the San Joaquin Valley, all <3s, all <10s, and all values between 3 and 10 are assigned the same value below 10, resulting in tied ranks for all, and the rank-sum test performed. Censoring data in this way uses the same information portrayed in the censored boxplot of Figure 9.1, where only the frequencies of observations below 10 in each group, and not the values themselves, are used to compute the positions of the box. Each of the 20 observations (30 percent) in the Alluvial Fan zone below 10 µg/L, and the 12 observations (24 percent) in the Basin Trough zone below 10, is assigned a rank of 16.5, the median of numbers 1 through 32. Then the Mann-Whitney test is computed. It is essential that software be able to compute tie corrections to test statistics for the many tied values resulting from nondetects.

Results of the Mann-Whitney test on the zinc data censored in this way are below. The test finds a definite and reliable difference in the distributions of zinc concentration between the two zones (p = 0.0185). Median concentrations are higher in the Basin Trough zone. No false 'information' was fabricated as was done with substitution. No signal was imposed by substitution that was not there, and any signal present (including the proportions of values below 10 in the two groups) is efficiently extracted from the data. As compared to MLE methods, no distribution was assumed for the data.

```
Mann-Whitney test for ZnMaxDL

Alluv_Zn   N =  67      Median =        10.00
Basin_Zn   N =  50      Median =        18.50
Point estimate for ETA1-ETA2 is        -5.00
95.0 Percent CI for ETA1-ETA2 is  (-10.00,-0.00)
W = 3533.5
Test of ETA1 = ETA2 vs ETA1 not = ETA2 is significant at 0.0210
The test is significant at 0.0185 (adjusted for ties)
```

A standard nonparametric test can always be computed on censored environmental data by censoring all values below the highest detection limit to a common value. The advantage of this procedure is that it uses standard software. The disadvantage is that it loses information in comparison to score tests, described below, and so has less power than those more complicated procedures. However, when there are sufficient differences between groups, this simple method may provide all the power needed to reject the hypothesis of no difference. As with other nonparametric methods, the assumptions of a normal distribution and equal variance are not necessary. These assumptions are difficult to check with censored data, as the entire distribution of data cannot be determined.

*Multiple detection limits. Method II – Score tests*

Score tests are nonparametric tests that determine whether distribution functions differ among groups of censored data. The distribution function of each group is called its survival function. Some score tests are direct extensions of the rank-sum test, including the "generalized Wilcoxon test", the "Peto-Prentice test", and the "Gehan test". If uncensored data are input to these Wilcoxon-type tests, the results are identical to the Wilcoxon rank-sum (Mann-Whitney) test. Score tests were designed to handle data censored at multiple detection limits, using the information contained in detected values between detection limits in addition to the information in the proportion of values below each detection limit. The survival function for left-censored environmental data estimates the cumulative distribution function i/n, where values for i are the ranks of data from smallest to largest. For values below a detection threshold, the survival probability i/n cannot be calculated exactly because the ranks of these values are not known. Score tests assign an estimated rank, or score, to each censored value.

*The Gehan test*

To illustrate exactly how a score test works, the Gehan test statistic (Gehan, 1965) will be manually computed for cadmium concentrations in fish livers (Cd.xls) in the Southern Rocky Mountain and Colorado Plateau physiographic regions. This data set is small and therefore easy to use for demonstrating how the Gehan test works. The rocks of the Southern Rocky Mountains are more mineralized, and contain considerably more trace metals including cadmium, than do rocks of the Colorado Plateau. Of interest is whether fish livers show this same pattern, indicating that concentrations in the host rock show up in stream waters, and subsequently in the biota of the two regions. The null hypothesis is that there is no difference in cadmium concentrations in livers of fish from the two regions. The alternative hypothesis is that concentrations in fish from the Southern Rocky Mountains are higher than those from the Colorado Plateau, a one-sided test.

There are 9 observations from the Colorado Plateau region, and 10 observations in the Southern Rocky Mountain region. The data are listed in the top row and left column of Table 9.2.

For the $n=9$ values $x_1$ to $x_n$ in the Colorado Plateau, and the $m=10$ values $y_1$ to $y_m$ in the Southern Rocky Mountains, there are $n \cdot m = 90$ possible comparisons. These comparisons, shown in the center cells of the table, are called $U_{ij}$, where

$$U_{ij} = \quad \begin{array}{ll} -1 & \text{for } x_i > y_j \ (y_j \text{ may be a nondetect}) \\ +1 & \text{for } x_i < y_j \ (x_i \text{ may be a nondetect}), \text{ and} \\ 0 & \text{for } x_i = y_j \quad \text{or} \\ & \text{for indeterminate comparisons } (<10 \text{ to a } 5). \end{array} \quad (9.2)$$

**Table 9.2.** Comparison of cadmium concentrations in fish livers of the Rocky Mountain region.

| S. Rocky Mts. | 81.3 | 4.9 | 4.6 | 3.5 | 3.4 | 3 | 2.9 | 0.6 | 0.6 | <0.2 |
|---|---|---|---|---|---|---|---|---|---|---|
| CO Plateau | | | | | | | | | | |
| 1.4 | 1 | 1 | 1 | 1 | 1 | 1 | 1 | −1 | −1 | −1 |
| 0.8 | 1 | 1 | 1 | 1 | 1 | 1 | 1 | −1 | −1 | −1 |
| 0.7 | 1 | 1 | 1 | 1 | 1 | 1 | 1 | −1 | −1 | −1 |
| <0.6 | 1 | 1 | 1 | 1 | 1 | 1 | 1 | 1 | 1 | 0 |
| 0.4 | 1 | 1 | 1 | 1 | 1 | 1 | 1 | 1 | 1 | −1 |
| 0.4 | 1 | 1 | 1 | 1 | 1 | 1 | 1 | 1 | 1 | −1 |
| 0.4 | 1 | 1 | 1 | 1 | 1 | 1 | 1 | 1 | 1 | −1 |
| <0.4 | 1 | 1 | 1 | 1 | 1 | 1 | 1 | 1 | 1 | 0 |
| <0.3 | 1 | 1 | 1 | 1 | 1 | 1 | 1 | 1 | 1 | 0 |

The Gehan test statistic W is the sum of the values for $U_{ij}$, or

$$W = \sum_{i=}^{n} \sum_{j=}^{m} U_{ij} \qquad (9.3)$$

For Table 9.2 there are 75 +1s and 12 −1s, so W = 63. If the null hypothesis were true, approximately half of these comparisons would be positive, and half negative, and W would be something close to 0. After computing W, it is standardized by dividing by a measure of its standard error (the square root of the variance),

$$Z = \frac{W}{\sqrt{Var[W]}} \qquad (9.4)$$

and the ratio Z is compared to a standard normal distribution. The equation for the variance of W must take into account the possibly large number of ties. When the two distributions are not equal, U, W, and Z are large in absolute value, and the null hypothesis is rejected. Note that if there were no censoring, all n•m comparisons could be quantified, and U would equal the rank-sum test statistic.

Two alternative formulae for the variance are common, the permutation and the hypergeometric estimates. Most software uses the hypergeometric variance. Permutation estimates are simpler, but are invalid when the censoring rate differs between the groups (one group has generally higher detection limits than the other). The permutation variance of W is:

$$Var[W] = \frac{mn \sum h^2}{(m+n)(m+n-1)} \qquad (9.6)$$

In this equation, the sample sizes m and n are the numbers of observations in the two groups. $h^2$ is the sum of squared values for h, a counting statistic related to the ranks of the data. To compute h (Table 9.3), count the number of observations known to be greater than each observation (G), and the number known to be less than each observation (L), when the data are sorted in ascending order (see Table 9.3). The difference between these two numbers equals h,

**Table 9.3.** Calculations for the variance of the Gehan test statistic

| Cd | G | L | h | h2 |
|---|---|---|---|---|
| < 0.2 | 15 | 0 | 15 | 225 |
| < 0.3 | 15 | 0 | 15 | 225 |
| < 0.4 | 15 | 0 | 15 | 225 |
| 0.4 | 12 | 3 | 9 | 81 |
| 0.4 | 12 | 3 | 9 | 81 |
| 0.4 | 12 | 3 | 9 | 81 |
| < 0.6 | 12 | 0 | 12 | 144 |
| 0.6 | 10 | 7 | 3 | 9 |
| 0.6 | 10 | 7 | 3 | 9 |
| 0.7 | 9 | 9 | 0 | 0 |
| 0.8 | 8 | 10 | −2 | 4 |
| 1.4 | 7 | 11 | −4 | 16 |
| 2.9 | 6 | 12 | −6 | 36 |
| 3 | 5 | 13 | −8 | 64 |
| 3.4 | 4 | 14 | −10 | 100 |
| 3.5 | 3 | 15 | −12 | 144 |
| 4.6 | 2 | 16 | −14 | 196 |
| 4.9 | 1 | 17 | −16 | 256 |
| 81.3 | 0 | 18 | −18 | 324 |
| | | | $\Sigma =$ | 2220 |

$$h = G - L \qquad (9.7)$$

which when squared and summed over all data, equals $\sum h^2$. For the cadmium data, $\sum h^2$ equals 2220. The variance of W is $\frac{9 \cdot 10 \cdot 2220}{(19)(18)} = 584.21$. The standard error of W is the square root of the variance, or 24.17 .

The Gehan test statistic Z is therefore

$$Z = \frac{63}{24.17} = 2.61$$

producing a p-value from the standard normal distribution = 0.0045. The conclusion is therefore to reject the null hypothesis of equality, finding that cadmium concentrations in fish livers are higher in the Southern Rocky Mountains than in the Colorado Plateau.

*The generalized Wilcoxon test*

Peto and Peto (1972) proposed a modification to the Gehan test called the "generalized Wilcoxon test". Prentice (1978) and Prentice and Marek (1979) elaborated on its properties, so the test is also called the Peto-Prentice test. Scores for the generalized Wilcoxon test are a weighted version of the Gehan test, adjusting the U scores of +1 or –1 by the survival function (edf) at that observation to create a new score. The U score for the generalized Wilcoxon test is

$$U_{ij} = \quad \begin{array}{ll} S(t_i) + S(t_{i-1}) - 1 & \text{for all uncensored observations } t \\ S(t_{i-1}) - 1 & \text{for all censored observations } t^* \end{array} \qquad (9.8)$$

where $S(t_{i-1})$ is the value of the survival function for the previous uncensored observation. For the first observation in the data set i=1, and the value of $S(t_0)$ equals 1. There is a 100 percent probability of exceeding a value smaller than the smallest observation in the data set.

The scores for one group are summed to obtain the test statistic W:

$$W = \sum_{i=1}^{n} U_i \qquad (9.9)$$

Dividing W by the square root of the variance for this statistic produces a Z statistic that can be compared to a table of the standard normal distribution. The permutation variance of W is

$$\text{Var}[W] = \frac{mn\sum U^2}{(m+n)(m+n-1)} \qquad (9.10)$$

where the sample sizes m and n are as before.

To continue the cadmium in fish livers example, computations for the generalized Wilcoxon test are listed in Table 9.4. Scores for the S. Rocky Mountain region are summed to produce the test statistic. If the score for the Colorado Plateau had been selected instead, U would be the same magnitude, with the opposite sign.

Flipping the Cd data into a right-censored variable (the "FlipCd" column) produces values that look like t or "time to censoring" of traditional survival analysis. The "Number Beyond" column lists the number of observations known to exceed the value of t. This is the same as the number of observations below the original cadmium concentrations, and so equals the ranks of the cadmium observations minus one. The survival function S(t) is the probability of surviving beyond each observation of FlipCd. This survival function is identical to the empirical distribution function of the original data, and equals i/n, where i is the rank of the original observations from low to high. Here tied observations were assigned tied ranks, as is standard in hypothesis testing. For example, the three detected observations at a cadmium concentration of 0.4 would have had the ranks of 3, 4 and 5 had there been enough precision in the measurements to tell the observations apart. Without that precision, any of the three observations could be the highest, or lowest. All three are therefore given a rank of 4, the median of the three possible ranks. In the survival analysis literature, tied values often follow another convention, assigning the minimum value for S, rather than the median value used here. Using the median assures that the sum of ranks for data with ties is the same as it would have been without ties, an important property for hypothesis tests.

If the null hypothesis is true, observations for each group will be randomly scattered through the list in Table 9.4, with about half of the scores positive and half negative. So W, the sum of the scores, will be near zero. If the null hypothesis is not true, the data from one group will be predominately near the top, or the bottom, of the list in Table 9.4. Consequently the absolute value of W will be larger than zero. From Table 9.4, the test statistic Z equals 2.637, and from a table of the standard normal distribution the associated one-sided p-value is 0.0042. The null hypothesis is soundly rejected, and it is concluded that cadmium concentrations in fish livers in the Southern Rocky Mountains are higher than those in fish from streams in the Colorado Plateau.

Using the right-censored flipped data, Minitab's® survival analysis software computes the generalized Wilcoxon test for the cadmium data. Notice that the test statistic for the test is 7.17 (see below), larger than the value calculated by hand. Minitab® uses an alternate form of the test statistic that follows a chi-square distribution rather than the normal distribution. The value of the chi-square test statistic will approximately equal the square of the test statistic computed using the normal approximation. For the cadmium data, $2.637^2$ equals 6.95, close to the 7.17 produced by Minitab®. The p-values for the two forms of the test should be approxi-

**Table 9.4.** Computation of the generalized Wilcoxon test for the cadmium data.

| Cd | CENSORING | REGION | FlipCd | Number Beyond | S(t) | U |
|----|-----------|--------|--------|---------------|------|---|
| 81.3 | 1 | S RKY MT | 18.7 | 18 | 0.947 | 0.947 |
| 4.9 | 1 | S RKY MT | 95.1 | 17 | 0.895 | 0.842 |
| 4.6 | 1 | S RKY MT | 95.4 | 16 | 0.842 | 0.737 |
| 3.5 | 1 | S RKY MT | 96.5 | 15 | 0.789 | 0.632 |
| 3.4 | 1 | S RKY MT | 96.6 | 14 | 0.737 | 0.526 |
| 3 | 1 | S RKY MT | 97 | 13 | 0.684 | 0.421 |
| 2.9 | 1 | S RKY MT | 97.1 | 12 | 0.632 | 0.316 |
| 1.4 | 1 | COLO PLT | 98.6 | 11 | 0.579 | 0.211 |
| 0.8 | 1 | COLO PLT | 99.2 | 10 | 0.526 | 0.105 |
| 0.7 | 1 | COLO PLT | 99.3 | 9 | 0.474 | 0.000 |
| 0.6 | 1 | S RKY MT | 99.4 | 7 | 0.368 | −0.158 |
| 0.6 | 1 | S RKY MT | 99.4 | 7 | 0.368 | −0.158 |
| 0.6 | 0 | COLO PLT | 99.4 | 7 | 0.000 | −0.632 |
| 0.4 | 1 | COLO PLT | 99.6 | 4 | 0.211 | −0.421 |
| 0.4 | 1 | COLO PLT | 99.6 | 4 | 0.211 | −0.421 |
| 0.4 | 1 | COLO PLT | 99.6 | 4 | 0.211 | −0.421 |
| 0.4 | 0 | COLO PLT | 99.6 | 2 | 0.000 | −0.789 |
| 0.3 | 0 | COLO PLT | 99.7 | 1 | 0.000 | −0.789 |
| 0.2 | 0 | S RKY MT | 99.8 | 0 | 0.000 | −0.789 |
| | | | W(S RKY MT) = | | | 3.316 |
| | | | Var(S RKY MT) = | | | 1.581 |
| | | | Z = | | | 2.637 |

mately the same. Note that the generalized Wilcoxon test can also be used to compare three or more distributions, analogous to the Kruskal-Wallis test. Therefore the p-values for the chi-square test statistic in Minitab® are always two-sided, as they would be for a Kruskal-Wallis test. To obtain a one-sided p-value when comparing two groups, first check that the observed differences are in the same direction as expected by the alternate hypothesis. If so, divide the reported p-value (0.0074) by two to obtain the one-sided p-value (0.0037). This value is very similar to the significance level of 0.0042 previously computed by hand.

```
Comparison of Survival Curves

Test Statistics
Method       Chi-Square    DF      P-Value
Log-Rank       5.5260      1       0.0187
Wilcoxon       7.1707      1       0.0074
```

*Score test for the Zn data from two geologic zones*
To test the multiply-censored zinc data of Millard and Deverel (1988) for differences between geologic zones, each Zn observation is subtracted from a large number in order to produce right-censored data. This large number could be any value larger than the maximum Zn observation. The maximum observation is 620 µg/L, so 623 is arbitrarily chosen. The flipped data are stored in a new column labeled 'FlipZn', where FlipZn = 623 − Zn.

The generalized Wilcoxon score test determines whether the survival distribution of FlipZn is the same in the two zones. The test is therefore determining whether there are differences in the cumulative distributions of the original data. If the p-value for the test is small, the distribution of zinc concentrations differs between the groups. The test is computed in Minitab® using the command:

Stat > Reliability/Survival > Distribution Analysis (Right censoring) > Nonparametric Distribution Analysis

where the grouping variable 'Zone' is entered in the 'By variable' dialog box. The procedure results in the following output:

```
Distribution Analysis: FlipZn by Zone
Comparison of Survival Curves

Test Statistics
Method       Chi-Square    DF      P-Value
Log-Rank      2.84260      1       0.092
Wilcoxon      5.54396      1       0.019
```

The Wilcoxon test statistic is 5.54, and the two-sided p-value is 0.019. The null hypothesis that the two groups have zinc concentrations with the same distribution is rejected. Note that the p-value is essentially the same as for the rank-sum test on data censored at the highest detection limit, reported earlier in this chapter. The additional information in these data attributable to the arrangement of detected values below 10 µg/L is small, compared to the information in the proportion of observations in each group above and below 10, and in the values for observations above 10 in each group. In other data sets the information contained in the multiply-censored pattern might be crucial to detecting differences, producing a p-value for the score test considerably smaller than for the one-threshold rank-sum test.

In Figure 9.5 the empirical distribution functions (survival functions) for the two groups are plotted. The vertical Percent scale tracks the estimated percentiles for each group (median = 0.5, etc.). The curves are similar at low concentrations (large flipped values), but are different for values around the 70th to 95th percentiles. For these percentiles the Alluvial Fan zone has generally larger flipped values, and thus lower Zn concentrations, than does the Basin Trough zone. This is consistent with what was seen in the boxplots of Figure 9.1.

Estimates of mean, median and interquartile range are given by Minitab® for the flipped data in each group. Location estimates (mean, median) must be re-scaled to concentration units by subtracting them from the large constant used to flip the data. This has been done and is presented in Table 9.5. Estimates of variability (standard deviation, variance, interquartile range) do not need to be re-scaled; they are the same on both the flipped and original scales.

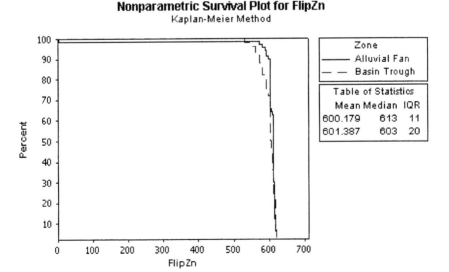

**Figure 9.5.** Survival plot of flipped zinc concentrations. Higher values of 'flipZn' (flipped data) correspond to lower concentrations.

**Table 9.5.** Summary statistics (Kaplan-Meier estimates) for Zn concentrations in the two geologic zones. Statistics for the flipped data are in parentheses.

|              | Mean          | Median    | IQR      |
|--------------|---------------|-----------|----------|
| Alluvial Fan | 22.82 (600.18)| 10 (613)  | 11 (11)  |
| Basin Trough | 21.61 (601.39)| 20 (603)  | 20 (20)  |

*Comparisons among score tests*

Latta (1981) evaluated the power of a number of two-sample score tests under conditions which included unequal sample sizes in the two groups, and unequal censoring (mix of detection limits) between the groups. The latter is a particularly difficult trait to overcome, because the pattern produced by unequal levels of censoring appears to be a signal instead of a design flaw. As seen in Chapter 1, this is a particularly severe problem for substitution methods.

Gehan, generalized Wilcoxon, logrank and other score tests were compared in Latta's large Monte Carlo experiment. Several versions of the test statistic variance (denominator of the standardized test statistic) were used, including asymptotic variance and permutation variance estimates. Of these tests, the Gehan and generalized Wilcoxon (Peto-Prentice) tests exhibited the most power when the underlying data were lognormal, the distribution most often used to model environmental data. The test with the overall best performance, including being able to accommodate unequal sample sizes and some measure of unequal censoring mechanisms, was the generalized Wilcoxon test using the asymptotic variance estimate. Environmental scientists would do well to look for software performing this version of a score test. In passing, Latta noted that for uncensored data with a lognormal shape, normal-scores tests have greater power than other nonparametric tests.

Millard and Deverel (1988) picked up on Latta's suggestion of normal-scores tests, recognizing that water-quality data often exhibit shapes similar to a lognormal distribution. They employed a normal-scores test based on Peto-Prentice type scores (Prentice and Marek, 1979) with a permutation variance estimate, and found it to be the "best behaved" score test for censored lognormal data. This tests produces a p-value of 0.02 (rounded) for the zinc data used in this chapter, so its results were essentially identical to those of the generalized Wilcoxon test reported above. Unfortunately, this censored normal-scores test is not available in most commercial statistical software. As of this writing, only the S-Plus add-on module for Environmental Stats (Millard and Neerchal, 2001), written by Millard, will perform this test. When using other statistical software, look for the Peto-Prentice or generalized Wilcoxon tests, to achieve high power for multiply-censored environmental data that are shaped something close to a lognormal distribution.

**Value of the information in nondetects**

Some scientists have felt that nondetects carry no information, and so the fewer the detected values in a data set, the less information is present. This is not correct. To illustrate, Zn concentrations in the Alluvial Fan zone, the zone with lower median, were altered by changing some of the detected values above 10 to a <10, and some of the <10s to <3s. The result is that the Alluvial Fan zone contains 54% nondetects, while the Basin Trough zone remains at 24% nondetects. The overall effect is to lower the zinc distribution in the Alluvial Fan zone, while increasing the proportion of nondetects. The signal in the data should be stronger, and should result in a lower p-value than for the original Zn concentrations, even though there are fewer detected values. Using the generalized Wilcoxon test on the altered data, the resulting p-value is $< 0.001$, more than an order of magnitude lower than the p-value of 0.019 for the unaltered data.

```
Distribution Analysis: FlipZn2 by Zone
Comparison of Survival Curves

Test Statistics
Method        Chi-Square   DF   P-Value
Log-Rank        17.6936    1    0.000
Wilcoxon        14.4449    1    0.000
```

The increased separation in the altered data is seen in their survival functions (Figure 9.6), where flipped data for the Alluvial Fan zone are now higher than in the original data, and more separate from the Basin Trough zone data. Higher flipped concentrations result from lower Zn concentrations in the original units.

Several guidance documents have recommended that statistical tests not be run with data having a high proportion of nondetects. For example, USEPA (2002b, p. A-9) recommends that if there are more than 40 percent nondetects in any group, the rank-sum test should not be used. There appears to be little justification for setting these types of rules. If the proportion of nondetects is similar in the two groups, the weight of evidence favors the null hypothesis. If the proportion differs significantly, as with 54% versus 24% nondetects for the altered zinc data, the null hypothesis will likely be rejected. In Chapter 10, TCE concentrations averaging 80% nondetects can be differentiated among three groups using the Kruskal-Wallis test, the multi-group equivalent of the rank-sum test. Tests that efficiently extract information from censored data, such as Wilcoxon score tests, will respond to the information contained in the data. There is no need to limit their use.

**Transformations with score tests**

Like other nonparametric tests, score test results are invariant when applied to data transformed using power functions such as the square root, logarithm, or inverse. Thus there is little reason to use transformations when performing score tests.

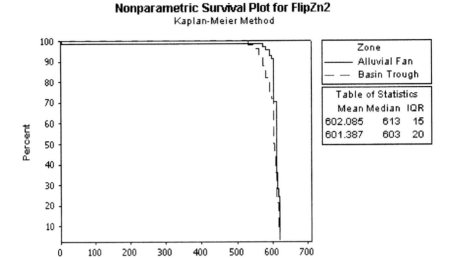

**Figure 9.6.** Survival functions of the altered Zn data with increased proportion of nondetects in the Alluvial Fan zone. Compare with Figure 9.5.

However, one possible reason to transform data prior to a score test is to produce survival plots where the differences between groups are more easily seen. Data with high outliers produce survival functions having a long left-hand 'tail', as seen in the Zn concentrations of Figure 9.5. This is a result of the right skewness for these data first seen in the boxplots of Figure 9.1.

Transformations to alter the shape of the data are performed first, prior to flipping the data to a right-censored scale. For the zinc example, Zn concentrations were log-transformed, the logs flipped, and the generalized Wilcoxon score test performed. The results below show that an identical p-value is produced for the original and log-transformed Zn data (0.019). There is no need for this transformation strictly from the viewpoint of the test itself. However, the survival functions of the two groups look different, as seen in Figure 9.7. The logarithms are more symmetric than in the original units, decreasing the length of the left-hand 'tail' in the survival functions, and drawing attention more clearly to the differences in concentration between the two zones.

```
Distribution Analysis: FliplnZn by Zone
Comparison of Survival Curves

Test Statistics
Method       Chi-Square   DF    P-Value
Log-Rank     2.84260      1     0.092
Wilcoxon     5.54396      1     0.019
```

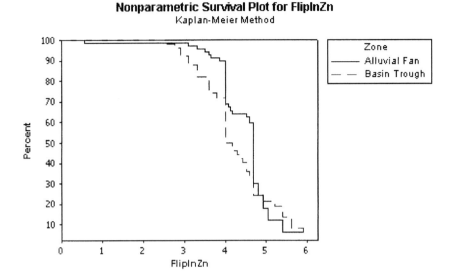

**Figure 9.7.** Survival functions of the logs of Zn concentrations. Compare with Figure 9.5.

Power transformations with coefficients less than 0, such as the inverse transformation $1/x$, change both the shape of the data as well as flipping its direction. Therefore, using $1/x$ may be a one-step process for making right-skewed data more symmetric, as well as turning a left-censored variable into a right-censored variable. Use of power functions like $1/x$ will not change the resulting p-values of nonparametric score tests (note that p-values of parametric tests will change as the shape of the distribution changes). However, estimates of mean, median and other statistics will be in different units than are the original data. Using any power transformation, log or $1/x$ included, requires retransforming estimates of summary statistics back into original units. The retransformation of power functions may make interpretations difficult (for more on transformation bias see Helsel and Hirsch, 2002). The process of 'flipping' the data used in this book, a linear transformation that does not alter the distance between observations, does not alter the data shape. No power transformation is involved, so that differences between the means or medians of flipped data are identical to differences in the means or medians of the original data. Interpretations of flipped data are simpler than for retransformations of power-transformed data.

## Paired observations

A variation on the two-group design occurs when observations in each group are purposely paired with one another to block out sources of background noise and focus on the effect being studied (Helsel and Hirsch, 2002, Chapter 6). With this structure, both groups have the same number of observations, and the first observation in the first group is linked to the first observation in the second group. Similarly, the second observation in the first group is linked to the second observation in the second group, the third with the third, and so on. Observations in the first group may be thought of as the 'starting point', and in the second group as the 'ending point'. The test determines if there are differences between the starting and ending points, even though the starting points may differ from pair to pair.

For uncensored data the standard tests for this design are the one-sample (or paired) t-test, and the nonparametric signed-rank test. Differences between each pair of observations are tested to see if their mean (t-test) or median (signed-rank test) difference is significantly different from zero. An alternate nonparametric test sometimes used with paired observations is the sign test. The sign test does not compute the magnitude of differences between pairs of observations, but records only whether there is an increase or decrease between the two values. This test determines if the proportion of increases, or decreases, is significantly different than the expected frequency of 50 percent. Not requiring an estimate of the magnitude of difference makes the sign test very useful for censored data.

With censored data the differences between pairs having one or more nondetects cannot be determined exactly. The same three options are open to the scientist for testing differences between paired groups – substitution, maximum likelihood, and nonparametric methods. As with the other examples in this book, substitution is fraught with problems. Results will vary with the values substituted. Substitution assumes that more is known about the data than is actually known. With multiple detection limits, the chance for disaster increases. No more time will be spent here on how substitution might be used for paired data. It is best avoided.

*Maximum likelihood estimation for paired data*

With maximum likelihood estimation, the mean difference between the starting and ending points is tested to determine if it is significantly different from zero. To do this, a confidence interval is constructed around the mean difference using MLE. If the confidence interval does not include zero, the differences are declared nonzero, and the two columns of data are declared to be different.

For censored data, differences must be calculated as an interval; the differences are interval-censored data. MLE can be used to compute the mean and standard deviation, and therefore a t-interval around the mean difference, for interval-censored data. For paired data, MLE requires an assumption that the paired differences follow a specified distribution. This assumption should first be checked before accepting the results of the MLE procedure.

As an example of paired data, atrazine concentrations in groundwater were measured at 24 wells in June, and at the same wells again in September, to determine whether concentrations had increased due to application of atrazine at the surface (Junk et al., 1980). Well to well differences are not of concern, and are 'blocked out' by the pairing process. All that is of interest is determining whether concentrations have increased for any given well during the time period. Several values are censored at the single detection limit of 0.01 µg/L. The data are found in atra.xls and shown in Table 9.6.

To compute a t-test for whether zero is included in the confidence interval around the mean difference, the differences are calculated (column 3 of Table 9.6). For nondetects there is a range of differences; the lowest and highest possible differences for the pair are stored separately in two columns. For pairs of detected observations, the same value for the difference is entered in each column. The two columns become the interval endpoints columns for MLE, with the column of smallest differences designated the Start column and the largesst differences the End column. The actual difference is somewhere within that interval. Using maximum likelihood, the mean difference and its 95% confidence interval are estimated; the output appears below. If data follow a normal distribution, this procedure is the equivalent for censored data to the paired t-test. The Minitab® macro PMLE.mac computes the confidence interval around the mean, and reports the p-value for the null hypothesis that the mean difference equals zero:

```
Location parameter for

Variable       Lower    Estimate   Upper   ----+-----+-----+----
Sept - June   -3.119     3.927    10.97   (---------*---------)
                                          ----+-----+-----+----
                                          0.0      4.0    8.0
Test for Location Equal to 0

Chi-Square     DF        P
 1.19336        1      0.275
```

The 95% confidence interval extends from –3.1 to a 10.97, which includes zero. Therefore the mean difference Sept-June is not significantly different from zero at an $\alpha = 0.05$. For the test of whether the mean equals zero, the p-value of 0.275 also indicates that no significant difference from zero was found.

These data are severely non-normal, as shown by a probability plot of the differences (Figure 9.8). An assumption of a lognormal distribution may be better than using the normal distribution. The logarithm of each month's data is calculated, and the differences in the logarithms computed and tested. The test for whether the mean difference in log units equals zero is identical to a test for whether the ratio of the two geometric means equals one. This test is appropriate as long as the geometric mean (median) rather than the arithmetic mean is an appropriate measure of the center. If the data must remain in original units, some accomodation for the one large outlier in the last row of Table 9.6, a point that inflates the standard error of the mean and therefore lengthens the confidence interval, must be made.

*Nonparametric methods for paired data*

The sign test determines whether paired values from one group generally are higher than the values from the other (Helsel and Hirsch, 2002, p. 138).

**Table 9.6.** Atrazine concentration pairs.  Do concentrations increase from June to September?

| June | Sept. | Sept–June | Sign of Difference |
|---|---|---|---|
| 0.38 | 2.66 | 2.28 | + |
| 0.04 | 0.63 | 0.59 | + |
| 0.01 | 0.59 | 0.58 to 0.59 | + |
| 0.03 | 0.05 | 0.02 | + |
| 0.03 | 0.84 | 0.81 | + |
| 0.05 | 0.58 | 0.53 | + |
| 0.02 | 0.02 | 0.00 | 0 |
| <0.01 | <0.01 | 0.00 to 0.01 | + |
| <0.01 | <0.01 | −0.01 to 0.01 | 0 |
| <0.01 | <0.01 | −0.01 to 0.01 | 0 |
| 0.11 | 0.09 | −0.02 | − |
| 0.09 | 0.31 | 0.22 | + |
| <0.01 | 0.02 | 0.01 to 0.02 | + |
| <0.01 | <0.01 | −0.01 to 0.01 | 0 |
| <0.01 | 0.5 | 0.49 to 0.50 | + |
| <0.01 | 0.03 | 0.02 to 0.03 | + |
| 0.02 | 0.09 | 0.07 | + |
| 0.03 | 0.06 | 0.03 | + |
| 0.02 | 0.03 | 0.01 | + |
| 0.02 | <0.01 | −0.01 to −0.02 | − |
| 0.05 | 0.10 | 0.05 | + |
| 0.03 | 0.25 | 0.22 | + |
| 0.05 | 0.03 | −0.02 | − |
| <0.01 | 88.36 | 88.35 to 88.36 | + |

Comparisons between paired observations are recorded only as an increase, a decrease, or a tie. These are shown in the "Sign of Difference" column in Table 9.6. Because the magnitude of the difference is not used, the sign test is directly applicable to paired censored observations. Due to its paired structure, the sign test can be performed whenever one detection limit is used per x-y pair. Though the test cannot evaluate a pair of observations (x,y) at (<1, <3), it can evaluate data where one pair is (<1, 10) and a second pair (<3, 5). Both increase from x to y. Therefore multiple detection limits can in limited fashion be incorporated into the sign test. Though more powerful and more complicated score tests have been developed for multiple censoring thresholds, the sign test remains the easiest test to employ for the situation of one detection limit per x-y pair.

Consider again the atrazine data of Table 9.6: 24 pairs of concentrations measured with a detection limit of 0.01 μg/L. The sign test determines whether the pattern of pluses and minuses differs from the expected frequency of 50% for each when the null hypothesis is true. This example is a one-sided test – the question was "is there an increase from June to September?" – so change in only one direction is of interest. The results below are the standard Minitab® output for the sign test, and

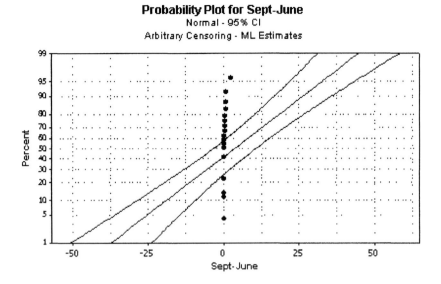

**Figure 9.8.** Probability plot of the paired atrazine differences Sept. – June.

indicates that there is indeed a difference, at a p-value of 0.0022. No substitutions or assumptions of a distributional shape for these data were required.

```
Sign Test for Median: S-J

Sign test of median = 0.00000 versus   >   0.00000

           N      Below    Equal    Above       P        Median
S-J        24       3        5        16      0.0022     0.02500
```

The sign test deals with ties by deleting the tied pairs from all calculations. For the atrazine example, the test computed the p-value as the likelihood of seeing 16 increases and 3 decreases out of a total of 19 pairs, ignoring the other 5 tied pairs. This may be acceptable for the small proportion of ties typically encountered in uncensored data, but for a larger proportion of ties it is not. Ignoring tied pairs will inflate the Type I error rate, artificially lowering the reported p-values (Fong et al., 2003). Data sets with a large proportion of ties should reflect their greater evidence for similarity than for data without tied values.

Fong et al. (2003) provide two methods for adjusting the sign test in the presence of tied pairs. Their "modified sign test" is implemented in the Minitab® macro Csign, producing p-values using the formula

$$p = \frac{\text{Prob}\left[N \geq \max\left(n^+, n^-\right)\right]}{\text{Prob}\left[N \geq \left\rangle\frac{n - n^o + 1}{2}\right\langle\right]}$$

(9.11)

where n = number of data pairs, $n^+$ = the number of increases, $n^-$ = the number of decreases, and $n^o$ = the number of ties. Capital N represents the binomial distribution with n observations evaluated at a the central 0.5 proportion, the probability of exceedance (p-values) for the sign test. The angle brackets in the denominator of equation 9.11 represent the floor function, so that $\rangle X \langle$ is the largest integer smaller than or equal to X. The modified p-value adjusted for ties is printed for the atrazine data by the Csign macro as:

```
p-value (adjusted for 'Equal' ties) = 0.0448
```

The five tied pairs out of a total of 24 pairs of data are evidence favoring the null hypothesis of equality, which when incorporated rather than deleted increase the p-value from 0.002 to 0.04. When computing the sign test for censored data, the modified sign test should be used rather than the default test computed by commercial statistical software so that ties, such as <dl versus <dl, are correctly incorporated into the test results.

*Multiple detection limits – the PPW test*

For multiple detection limits, score tests have been developed for the matched-pair design. These tests use a score statistic as a measure of the position of the observation in a data set. Scores are generally computed to be negative and positive, centered at zero on the midpoint of the data set. These scores are related to the ranks of the data by the equation

$$\text{Rank} = n \left[0.5 + 0.5 \cdot Score\right]$$

(9.12)

O'Brien and Fleming (1987) introduced the paired Prentice-Wilcoxon, or PPW, test. It uses the same form of scores used in the generalized Wilcoxon test for unpaired data, and is the standard test for the case of censored matched pairs. Akritas (1992) proposed an alternate test, performing a paired t-test on the ranks as defined above. Both tests have similar power (Akritas, 1992) for the situation of skewed data common in environmental studies. Neither the PPW nor the Akritas tests are found in commercial statistical software. However, a Minitab® macro (PPW.mac) to perform the Paired Prentice-Wilcoxon test is included on the web page that accompanies this book.

To compute the PPW test, the data are stacked into one column, and a Kaplan-Meier estimate of the survival function for the combined data is computed. Scores, the estimated percentiles of the survival function minus 0.5, are computed for each observation, both censored and uncensored. The scores are then split back up into their respective groups. If the null hypothesis is true and the two distributions are the same, differences between pairs of scores should be small, hovering around zero. In other words, the two paired observations should be located at similar places in the combined distribution; therefore their score values should be similar. If the distributions of the two groups differ, the paired observations will be located at different points of the combined survival distribution, with the scores from one data set consistently higher than the paired score from the other. The PPW test computes the differences between the paired scores, and determines whether the sum of these differences is significantly different from zero, using a normal approximation for the test statistic.

Four observations from the atrazine data of Table 9.6 were altered to produce paired data with two detection limits, at 0.01 and 0.05 µg/L. The altered data are shown in Table 9.7, with the altered observations at <0.05. Alterations were in the direction of stronger differences between the two months.

PPW scores for each pair are also shown in Table 9.7. The scores are computed on flipped data, so their signs are the reverse of what is expected from the original

**Table 9.7.** Atrazine concentration pairs from Table 9.6, altered to add a second detection limit.

| June | Sept | June Score S1 | Sept. Score S2 | June–Sept. Score S1 – S2 |
|------|------|------|------|------|
| 0.38 | 2.66 | −0.67 | −0.92 | 0.24 |
| < 0.05 | 0.63 | 0.38 | −0.84 | 1.22 |
| < 0.01 | 0.59 | 0.61 | −0.80 | 1.41 |
| 0.03 | 0.05 | −0.04 | −0.24 | 0.20 |
| 0.03 | 0.84 | −0.04 | −0.88 | 0.84 |
| 0.05 | 0.58 | −0.24 | −0.76 | 0.52 |
| 0.02 | 0.02 | 0.22 | 0.21 | 0 |
| < 0.01 | 0.01 | 0.61 | 0.61 | 0 |
| < 0.01 | < 0.01 | 0.61 | 0.61 | 0 |
| < 0.01 | < 0.01 | 0.61 | 0.61 | 0 |
| 0.11 | 0.09 | −0.55 | −0.43 | −0.12 |
| 0.09 | 0.31 | −0.43 | −0.63 | 0.20 |
| < 0.01 | 0.02 | 0.61 | 0.22 | 0.39 |
| < 0.01 | < 0.01 | 0.61 | 0.61 | 0 |
| < 0.01 | 0.50 | 0.61 | −0.71 | 1.32 |
| < 0.01 | 0.03 | 0.61 | −0.04 | 0.65 |
| 0.02 | 0.09 | 0.22 | −0.43 | 0.65 |
| < 0.05 | 0.06 | 0.38 | −0.35 | 0.73 |
| 0.02 | 0.03 | 0.22 | −0.04 | 0.26 |
| 0.02 | 0.01 | 0.22 | 0.61 | −0.39 |
| 0.05 | 0.10 | −0.24 | −0.51 | 0.27 |
| 0.03 | 0.25 | −0.04 | −0.59 | 0.55 |
| 0.05 | < 0.05 | −0.24 | 0.38 | −0.62 |
| < 0.01 | 88.36 | 0.61 | −0.96 | 1.57 |

observations. For uncensored observations:

> *Score* = 1 - 2*S*          where S is the Kaplan-Meier estimate for the percentile of the survival function.

For censored observations:

> *Score* = 1 - *Sj*          where Sj is the Kaplan-Meier estimate for the percentile of the survival function for the next smallest (in flipped units) uncensored observation.

Scores can be considered as scaled ranks of observations, although in the reverse order from the original units due to flipping the data. For example, the largest observation of 88.36 has the largest negative score, of –0.96. The second largest observation of 2.66 has the second largest negative score, of –0.92. All concentrations tied at a detection limit have scores identical to one another. Notice that the difference in scores equals 0 for pairs with tied data. As the number of ties increases, Z will get smaller, providing less evidence against the null hypothesis. Differences in scores are largest for pairs with large differences in ranks of the original observations. For the PPW test, the difference in the paired scores ($d = S_1 - S_2$) is tested using the test statistic

$$Z_{PPW} = \frac{\sum d}{\sqrt{\sum d^2}} \qquad (9.13)$$

by comparing Z to a table of the normal distribution. The output below shows that the September atrazine concentrations are significantly higher than their paired June concentrations (p=0.002). This is stronger evidence than that for the sign test, where the p-value was an order of magnitude larger. The larger differences between groups created when the data were altered is picked up by the PPW test's ability to incorporate multiple detection thresholds.

```
Paired Prentice-Wilcoxon test
(NonPar test for equality of paired left-censored data)

       Ho:   distribution of Sept = June

vs   Ha:   greater than

Test Statistic: 2.936
        p value: 0.002
```

*Multiple detection limits – the Akritas test*

For uncensored data, the nonparametric signed-rank test can be approximated by computing a paired t-test on the signed-ranks (Conover and Iman, 1981). The Akritas test (Akritas, 1992) extends this idea to data with censored values. Ranks of uncensored values are computed much as in the PPW test, using Kaplan-Meier statistics for the entire data set. Ranks of censored observations are computed by calculating Kaplan-Meier survival functions separately for each group and averaging the survival probabilities for the pair of observations that includes the censored value. The rank of the censored observation is computed from this averaged probability (Akritas, 1992). A paired t-test is then computed on the calculated ranks to evaluate the similarity of the set of paired observations.

As found by Conover and Iman (1981), rank-transform tests generally produce p-values slightly lower than those of the exact tests for the same situation. Null hypotheses are rejected a little more frequently than they should in simulation studies evaluating these tests. As a rank-transform test, this characteristic may also be true for the Akritas test. More detail on its computation is found in Akritas (1992).

## Comparing data to a standard using paired tests

Methods for paired observations can be used to determine whether the mean or median of one column of data equals or exceeds a fixed standard. The paired t-test, signed-rank test, and sign test are used with uncensored data to test compliance to a standard. For censored data, the MLE confidence interval and nonparametric PPW test can serve the same purpose, even for data with multiple detection limits.

Instead of two columns of data, comparisons to a standard are made by placing the value of the standard in every entry of the second data column. Differences are computed between the data and the standard, and the mean or median difference tested to determine whether or not it equals zero. Tests for compliance are usually set up assuming compliance; the alternate hypothesis is that the mean or median difference exceeds zero. If so, the mean or median of the column of data exceeds the standard.

As an example, consider whether the mean or median of the altered atrazine data for June of Table 9.7 exceeds a standard of 0.05 μg/L. There are two detection limits, at 0.01 and 0.05. Differences betwee the data and the standard are interval-censored values. The endpoints of the interval are identical when the data are above the reporting limit. Endpoints differ for censored observations, representing the range of possible differences. Using maximum likelihood (the %PMLE macro) and assuming a normal distribution, the mean concentration does not exceed the standard of 0.05. The chi-square test statistic is a two-sided test: for a one-sided test the p-value is 0.583/2, or 0.26. Also note that the lower 95% confidence bound extends below zero, so that zero is included as a possible estimate for the difference.

```
Test for Location Equal to 0
Chi-Square   DF       P
  0.300634    1    0.583
Bonferroni 95.0% (indiv 95.00%) Simultaneous Lower Bound
Location parameter for
  Variable      Lower      Estimate    -----+-----+-----+-----
June - Std    -0.03388    -0.008472    (-----------*---------
                                       -----+-----+-----+-----
                                       -0.024   -0.012   -0.000
```

Using the nonparametric PPW test, the median June atrazine concentration also does not exceed the standard of 0.05; the null hypothesis of equality of medians is not rejected (p = 1.0). Both the mean and median of these multiply-censored data are within the standard set for them.

```
Paired Prentice-Wilcoxon test
(NonPar test for equality of paired left-censored data)

     Ho:  distribution of June = Std
vs   Ha:  greater than

Test Statistic: -3.713
      p value:  1.000
```

### Summary of two-sample tests for censored data

Two-sample tests corresponding to the t-test, Mann-Whitney and sign tests are available for use with censored data. The decision whether to use a parametric or nonparametric test is made in the same way as for uncensored data, judging how closely the data, or their transformed values, follow a normal distribution. Censored t-tests are available in survival analysis by solving for regression parameters with maximum likelihood. The slope coefficient estimates the difference between the means of the two groups, and testing this coefficient determines whether the difference is significant. For nonparametric tests, the generalized Wilcoxon test expands on the uncensored Wilcoxon rank-sum (or Mann-Whitney) test, adjusting the ranks to incorporate information contained in nondetects. Wilcoxon tests can be used on both singly- and multiply-censored data. However, for a single detection limit these tests are identical to the standard rank-sum or sign test. Therefore standard nonparametric tests can be used for data with one detection limit.

Paired tests analagous to the paired t-test and signed-rank tests are available for multiply-censored data. They can also be used to test whether one set of multiply-censored data exceed a legal or other standard.

Substitution of values for nondetects may lead to serious errors, because signals not present in the data can be introduced or differences between groups can be obscured. Given the availability of methods that correctly incorporate the information in nondetects, both parametric and nonparametric methods, substitution methods for testing between groups should be totally avoided.

# Exercises

**9-1**     Eppinger et al. (2003) measured metals concentrations in stream sediments at 82 sites in New Mexico in 1996. After wildfires occurred throughout the region in 2000, each site was re-sampled to determine if concentrations had changed following the fires. Several mechanisms were proposed for why this might be so. Data for lead are found in SedPb.xls. Test to determine whether lead concentrations, some of which are recorded as below a single detection limit of 4 µg/L, have changed pre- and post-fire. Note that the data are paired by sampling location.

**9-2**     Yamaguchi et al. (2003) measured concentrations of the pesticide lindane in fish and eels collected at several sites in the United Kingdom. One site was below Swindon, an active industrial area draining to the Ray River, a tributary to the Thames. To avoid differences due to types of fish, data presented here are for only one species (Roach) at two sites, Swindon and a site further downstream on the Thames River, in the file Roach.xls. There was one detection limit, at 0.08 µg/kg. Test whether lindane concentrations are the same or different at the two sites, using both the parametric "t-test" performed with censored regression, and a nonparametric Wilcoxon score test.

**9-3**     Squillace et al. (1999) related VOC concentrations in groundwater throughout the United States to population density. The data for one compound, chloroform, in the state of California is presented in ChlfrmCA.xls. Observations are grouped by whether they are from urban areas (population density > 386 people per acre) or rural areas (population density < 386 people per acre). Determine if the mean/median concentration of chloroform is higher in the urban areas than in the rural areas. Use both a parametric and nonparametric test. There are two detection limits.

# 10

# COMPARING THREE OR MORE GROUPS

Scientists must at times evaluate environmental data that are classified into more than two groups. Are the means, or medians, or probabilities of observing a detected value the same within each group, or is at least one different? What can be said when there are nondetects below several reporting limits scattered throughout the groups? For uncensored data, comparisons among groups are made using the parametric Analysis Of Variance (AOV or ANOVA) and the nonparametric Kruskal-Wallis test. For data with nondetects, the methods surveyed in Chapter 9 for differentiating between two groups – substitution, maximum likelihood, and nonparametric methods – can be extended to three or more groups and employed here. Differences between means can be tested using maximum likelihood, using software for censored regression. Nonparametric Wilcoxon score tests look for differences in the cumulative distribution functions (or survival distributions) among the groups.

As the primary example for this chapter, concentrations of trichloroethylene (TCE) in shallow ground waters of Long Island, NY were reported by Eckhardt et al. (1989). The waters sampled were from wells surrounded by one of three land use types: low, medium, or high-density residential lands. Sources of TCE were expected in each land use type due to the past use of TCE as a septic system cleaner, and as a solvent in a variety of residential and light industrial uses. At issue is whether the occurrence of TCE is similar in waters under the three land use types, or whether concentrations appear to differ in at least one land use. Because samples were sent to different laboratories, and because precision changed over time, four detection limits at 1, 2, 4 and 5 µg/L are found in the data. Boxplots of the data are shown in Figure 10.1, where the boxes for all three land-use groups lie below the highest detection limit. The percentage of values above the highest detection limit of 5 µg/L in each of the three groups is listed in Table 10.1.

These data can be found in TCE.xls. Concentrations or detection limits are stored in the column named TCEConc. The column named BDL=1 indicates a censored observation with a value of 1, and a detected value with a 0. The land-use groups are listed in the column named Density.

**Substitution methods**

Substituting the detection limit for all nondetects can quickly be accomplished by ignoring the indicator variable, and using only the values stored in the concentration column. An analysis of variance (ANOVA) run directly on the concentration column produces an ANOVA with substitution of the detection limits. This produces a high bias for the means of each group. As already shown in previous chapters,

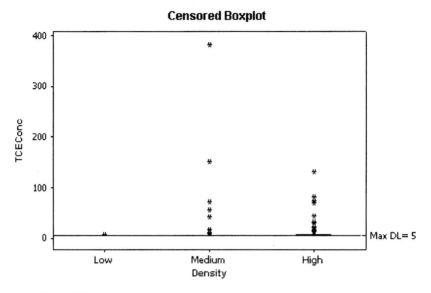

**Figure 10.1.** TCE concentrations in ground water under three land use types (Eckhardt et al., 1989).

**Table 10.1.** Percent of TCE concentrations in Long Island ground water greater than 5 mg/L. Data from Eckhardt et al., (1989).

| Land Use Density | Low | Medium | High |
|---|---|---|---|
| Percent above 5 µg/L | 0 | 9 | 20 |

substitution in the case of multiple detection limits can either artificially produce patterns not seen in the data themselves or obscure patterns that should be detected. ANOVA on the column of TCE concentrations produces the following results:

```
One-way Analysis of Variance

Analysis of Variance for TCEConc
Source      DF          SS          MS          F           P
Density      2        1120         560        0.60       0.547
Error      244      225876         926
Total      246      226996
```

ANOVA compares a ratio of the mean square (MS) for the effect (Density) to the MS of the background noise (Error). This ratio is the F statistic for the test, 560/926 = 0.60. The expected value for F when there is no differences among the group means is approximately 1. If the F statistic is much larger than 1, the null hypothesis of no difference among means can be rejected. For the TCE data, the F statistic of 0.60 is not large. The magnitude of difference in means it represents can be expected to occur about 55 percent of the time when the null hypothesis is true (p = 0.547). This is insufficient evidence to reject the null hypothesis, and therefore mean TCE concentrations for the three groups are considered to be similar by this test. This result could also be due, however, to either the non-normality of the data (censored data sets are often quite skewed) or to the inaccuracy of fabricating the pattern of concentrations by substitution.

Substitution of either one-half the detection limit or zero also result in insignificant F-tests. ANOVA on substituted values fails in each case to find differences in the mean concentration among groups. Given the large proportion of censoring this should not be a surprise – the mean is not a particularly useful parameter to summarize a highly skewed and highly censored data set.

It should be noted that the simplest nonparametric test for these data, a contingency table test presented in a later section, finds significant differences in the proportions of data above 5 μg/L among the three land-use groups. Substitution's failure to see any differences emphasizes again the weaknesses of the method. There is no need to use substitution. There are better ways.

## Maximum likelihood estimation

MLE can be used with censored data to perform hypothesis tests similar to Analysis of Variance. As with other parametric methods, the results will be valid if the data used for the test closely follow the distribution assumed by the MLE. When the data do not match the distribution assumed by MLE, hypothesis tests may have low power to discern differences present between groups.

In order to use MLE, the data are usually coded as interval-censored data in the Interval Endpoints format described in Chapter 3. Censored observations have a lower bound of 0 and upper bound of the reporting limit. Hypothesis tests will then recognize that data cannot go negative. Detected observations have the same value in each of the two endpoint columns. Alternatively, for skewed distributions the data (or their reporting limits) can be log-transformed and then flipped. In addition to accounting for skewness, this transformation allows software for right-censored data to be used to conduct the test.

### Likelihood-ratio tests

Parametric hypothesis tests for differences among groups in the means of censored data can be accomplished using censored regression methods available in survival analysis software. A likelihood-ratio test equivalent in purpose to ANOVA's F test determines whether there is a significant increase in the log-likelihood of data classified by group in comparison to the log-likelihood for data when unclassified (the null likelihood). In other words, does the classification by group explain a significant portion of the variation observed in the data? If so, the means of the groups are not all similar, and classification reflects that the location of at least one group is not identical to the others. When the log-likelihood statistic with classification is significantly greater (less negative) than that for no classification, the null hypothesis of no difference among means is rejected. The test statistic is the "−2 log-likelihood"

$$-2\log\text{-likelihood} = -2*(L_{null} - L_{groups})\qquad(10.1)$$

where L is the log-likelihood of each situation. Because these are logs of the likelihood, subtracting one from the other is equivalent to a ratio of likelihoods, hence the name "likelihood ratio test".

Some software packages compute both the likelihood L and the −2 log-likelihood, comparing the latter to a chi-square distribution with k-1 degrees of freedom, where k is the number of groups. It is a one-step process. Minitab® and other packages require separate commands to print the model and null likelihoods, requiring you to compute the test statistic using equation 10.1, and its p-value. However this is not an onerous task. Before the TCE data are used as an example, the first step in computing a likelihood ratio test is to represent the data in the interval endpoints format.

*Representing interval-censored data*

TCE concentrations have a lower bound of zero , as do most environmental data. Some statistical software allows entry of low and high endpoints for censored data, called "interval censored" or "arbitrary censored" data. Setting the lower bound at zero is important in accurately representing the possible values of the data and for returning correct values for coefficients and tests. If no lower bound is set (software would generally represent the lower end with an *), values are assumed to be able to go as low as minus infinity. This produces a low bias for estimates of the mean, high bias for standard deviation, and incorrect test results in parametric hypothesis tests.

Survival analysis software assumes that data are right-censored with no upper bound. This is not an issue with nonparametric methods – the highest observation simply has the highest rank. However, flipping left-censored environmental data and running MLE survival analysis procedures with no upper bound can produce incorrect results. If a normal distribution is assumed, a lower bound of zero *must* be represented. This can be done with software that allows interval censoring, such as Minitab® and SAS®. When using a lognormal distribution, however, the lower end of the logarithms can go to minus infinity because this translates into a lower bound of zero in original units. If available software only allows right-censored data, compute the logarithms of data, flip the logarithms by subtracting from any number larger than the maximum log value, and run the procedure on these right-censored values ($y_{flip}$ in equation 10.2).

$$y_{flip} = C - \ln(x)\qquad(10.2)$$

where C is any number larger than the maximum of ln(x). The result when re-transformed back into units of x will be correct for lognormal data with a lower bound of zero, the latter represented by the plus infinity value for $y_{flip}$. As most environmental data more closely follow a lognormal rather than normal distribution, this procedure should work well for many data sets if interval-censored software is not available.

Using Minitab's® Regression with Life Data procedure, interval-censored data can be directly entered as the response variable in a censored regression (Figure 10.2). For the TCE data, the low end of the interval is the variable TCE0, where all nondetects are represented by a value of 0. The high end of the interval is the variable TCEConc, with nondetects set to the detection limit. The format is the "interval endpoints" format of Chapter 4. As shown later, censored regression software will produce a parametric test for differences among the means of these grouped data.

*Representing data groups with regression software*

Minitab® and several other survival analysis packages allow explanatory variables in a regression to be specified as a "factor" or "grouped variable" or a "category". These variables have a single value for every observation in a group and so define the group identity for each observation. These may be numeric or text variables – in the TCE example the group identifier is the text variable Density. Explanatory variables entered into the Model dialog box are designated as factors in Minitab® by also entering them into the Factors dialog box, as shown in Figure 10.2.

Specification of factor variables in regression can be accomplished by hand if the software does not have an option to do so. To do this, the group identifier is recoded into binary (0/1) variables, using the same process as for analysis of covariance with regression software (Helsel and Hirsch, 2002, Chapter 11). Membership in one of k groups is represented by (k-1) binary variables, each with values either as 0 or 1. Membership is usually indicated as the value 1. It takes two binary variables to represent the same information contained in the three groups designated by the variable Density. One variable is named MedEq1, and has a value of 1 for every observation in the Medium density group. Observations from the Low or High density groups have a value of 0 for the MedEq1 variable. Similarly, the variable HiEq1 has the value 1 for all data from the High density group, and zeros otherwise. The low density group is defined as the 'baseline' situation, with values for both MedEq1 and HiEq1 equal to 0.

*The TCE example*

With TCE concentrations in the interval zero to the detection limits represented by TCE0 and TCEConc, Density groups entered as a Factor (Figure 10.2), and the assumed distribution set to a normal distribution, Minitab's® MLE life-regression procedure will estimate an overall likelihood ratio used in determining whether the means of the three groups are significantly different. In essence it produces an

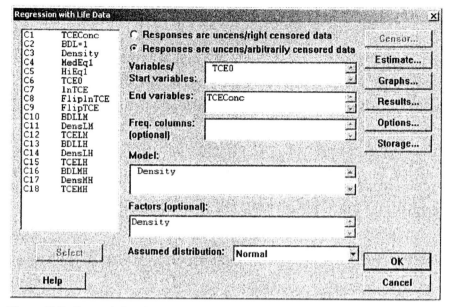

**Figure 10.2.** Entry window for censored regression in Minitab®. Interval ("arbitrary") censoring results from use of the Start and End variables.

ANOVA-type test where differences and their significance are estimated by maximum likelihood. The output of the procedure is:

```
                      Standard                    95.0% Normal CI
Predictor     Coef    Error      Z       P       Lower     Upper
Intercept   1.01988  6.08187   0.17    0.867   -10.9003   12.9401
Density
  High      6.74525  6.85866   0.98    0.325    -6.69748  20.1879
  Medium    7.00662  6.64111   1.06    0.291    -6.00971  20.0229
Scale      30.4048   1.36799                    27.8383   33.2078

Log-Likelihood = -1069.163
```

To determine whether significant differences in mean TCE are found between Density groups, the log-likelihood for no grouping (the null log-likelihood) is obtained by running the

Stat > Reliability/Survival > Distribution Analysis (Arbitrary Censoring) > Parametric Distribution Analysis

procedure, setting the assumed distribution to the same as for the grouping results, here a normal distribution. The output from the procedure estimates the mean and standard deviation for the entire TCEConc column without breaking it into groups, and produces the null log-likelihood:

|           |          | Standard | 95.0% Normal CI | |
|-----------|----------|----------|---------|---------|
| Parameter | Estimate | Error    | Lower   | Upper   |
| Mean      | 7.21985  | 1.93959  | 3.41830 | 11.0213 |
| StDev     | 30.4761  | 1.37120  | 27.9037 | 33.2857 |

Log-Likelihood = -1069.741

To test for whether there are significant differences in mean TCE among the three Density groups, compute the likelihood-ratio test statistic (equation 10.1)

$$-2 \log\text{-likelihood} = -2*(L_{null} - L_{groups}) = -2*(-1069.741 - [-1069.163]) = 1.156$$

This statistic is compared to a chi-square distribution with $(k-1) = 2$ degrees of freedom. The resulting p-value is 0.56, indicating that no significant differences are found among the three group means.

If there had been no designation for a grouping variable (factor) available in Minitab®, the two binary variables MedEq1 and HiEq1 described previously could have been entered as explanatory variables in the model dialog box (Figure 10.3). The resulting output shown below is identical to that produced when the factor designation was used. Internally the software designates all but one of the values for the factor variable as a 0/1 binary variable and enters those as explanatory variables in the regression model.

|           |         | Standard |      |       | 95.0% Normal CI | |
|-----------|---------|----------|------|-------|----------|---------|
| Predictor | Coef    | Error    | Z    | P     | Lower    | Upper   |
| Intercept | 1.01988 | 6.08187  | 0.17 | 0.867 | -10.9003 | 12.9401 |
| MedEq1    | 7.00662 | 6.64111  | 1.06 | 0.291 | -6.00971 | 20.0229 |
| HiEq1     | 6.74525 | 6.85866  | 0.98 | 0.325 | -6.69748 | 20.1879 |
| Scale     | 30.4048 | 1.36799  |      |       | 27.8383  | 33.2078 |

Log-Likelihood = -1069.163

*Importance of the assumed distribution*

The validity of results for a parametric method, whether involving censored data or not, depends on the adherence of the observed data to the assumed distribution. One of the common procedures for evaluating adherence to the assumed distribution is a probability plot of the residuals. Figure 10.4 is a probability plot of residuals from the censored regression for the TCE data. The departure of the residuals from the straight line indicates that it is unlikely they arose from a normal distribution. When data depart from the assumed distribution, the power of parametric procedures is low and findings of no difference between group means can result. Therefore the TCE data should be tested again, this time with a lognormal distribution.

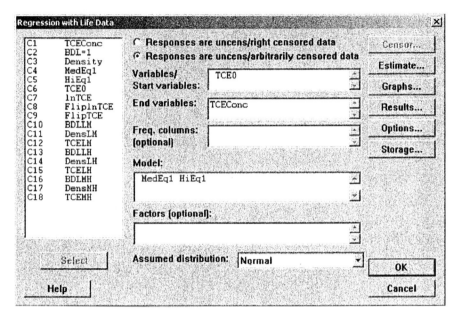

**Figure 10.3.** Entry window for censored regression using 0/1 binary variables instead of a factor.

Testing for differences between the mean logarithms of the three TCE groups can be quickly accomplished in Minitab® by choosing the lognormal distribution in the dialog box. The results below indicate that the groups' geometric means (medians) differ:

|           |          | Standard |        |       | 95.0% Normal CI |          |
|-----------|----------|----------|--------|-------|-----------------|----------|
| Predictor | Coef     | Error    | Z      | P     | Lower           | Upper    |
| Intercept | -3.78195 | 1.15396  | -3.28  | 0.001 | -6.04369        | -1.52021 |
| Density   |          |          |        |       |                 |          |
| High      | 3.05966  | 1.13781  | 2.69   | 0.007 | 0.829594        | 5.28974  |
| Medium    | 1.40339  | 1.11891  | 1.25   | 0.210 | -0.789640       | 3.59642  |
| Scale     | 2.85286  | 0.317678 |        |       | 2.29349         | 3.54867  |

Log-Likelihood = -308.700

The $-2\log\text{-likelihood} = -2*(L_{null} - L_{groups}) = -2*(-316.40 - [-308.70]) =$ 15.40, which when compared to a chi-square distribution with 2 degrees of freedom, corresponds to a p-value for the test of 0.0083. The probability plot of residuals (Figure 10.5) is much more linear than Figure 10.4. The data are more appropriately analyzed in log units.

How would this test be computed when software only allows right-censored data, and does not consider grouping variables? The same results would be obtained as above, assuming a lognormal distribution, using these steps:

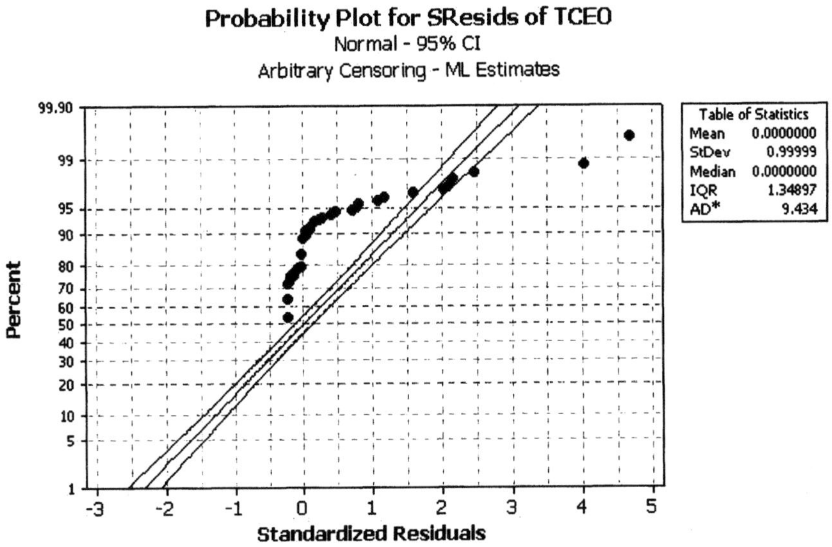

## Probability Plot for SResids of TCEO
### Normal - 95% CI
### Arbitrary Censoring - ML Estimates

**Figure 10.4.** Probability plot of residuals from the MLE test for differences in TCE group means. The data do not appear linear, falling outside of confidence bounds (solid lines) around the normal distribution.

1. Take logarithms of the TCE data.
2. Flip the logs by subtracting from a large constant. This produces a right-censored data set whose maximum at infinity will map back to a zero lower bound for TCE.
3. Create the two binary variables MedEq1 and HiEq1. Use these as the two explanatory variables in the regression equation. The entry window for the setup using Minitab® is given in Figure 10.6.
4. From the output (shown below) compare the log-likelihood of the model to the null log-likelihood. The test statistic value is the same 15.40 found for the more automated procedure above. Also note that the slopes for the two explanatory variables are just $(-1)$ times the values found above for the same variables. The sign is now negative because the data have been flipped.

```
                       Standard                    95.0% Normal CI
Predictor     Coef      Error      Z        P      Lower     Upper
Intercept  13.7819    1.15396   11.94    0.000    11.5201   16.0436
MedEq1     -1.40339   1.11891   -1.25    0.210    -3.59642   0.789640
HiEq1      -3.05966   1.13781   -2.69    0.007    -5.28974  -0.829594
Scale       2.85286   0.317678                     2.29349   3.54867

Log-Likelihood = -197.761
```

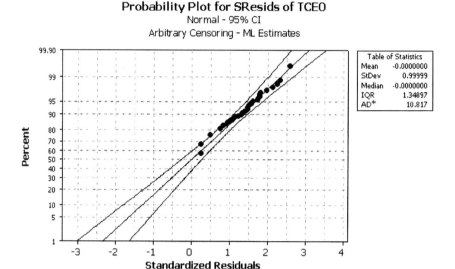

**Figure 10.5.** Probability plot of residuals from the MLE test for differences in TCE group means, assuming a lognormal distribution.

The $-2\log\text{-likelihood} = -2*(L_{null} - L_{groups}) = -2*(-205.46 - [-197.76]) =$ 15.40.

Either the automated grouping or the manual creation of binary variables results in the same likelihood-ratio test and p-value, the same individual slope coefficients (multiplied by $-1$ if the data were flipped), and the same tests of significance on the individual pairwise comparisons between groups. Pairwise comparisons are the way in which multiple comparisons can be computed using MLE software for censored data.

*Multiple Comparison Tests*

If an overall test is found to be significant, the next question is often "which groups differ from the others?" A series of individual comparisons between group means can be performed to answer this question. To determine the entire pattern of k group means requires $g=k(k-1)/2$ comparisons. For three groups, the g = 3 comparisons are between groups 1 and 2, groups 1 and 3, and groups 2 and 3. The end result might be something like "groups 1 and 2 are not significantly different, but both are lower than the mean of group 3". If an overall error rate of 5% is desired, so that there is no more than a 5% chance of making one error in the ordering of the means, each group comparison must be made at an individual error rate smaller than 5%. One commonly-used formula for determining the individual error rate is

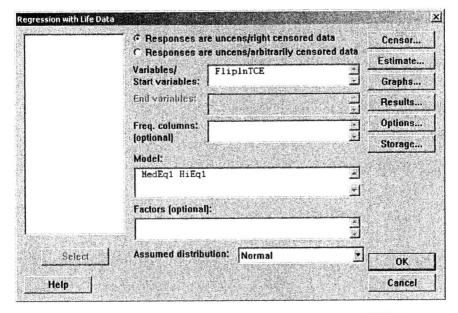

**Figure 10.6.** Entry window for right-censored flipped logarithms of TCE data.

Bonferroni's formula in equation 10.3:

$$\text{individual error rate} = \alpha / g \qquad (10.3)$$

where $\alpha$ is the desired overall error rate (often 5 percent, 0.05) and g is the number of comparisons between means to be made. Any p-values for tests on differences between two means that are below the Bonferroni-adjusted level are considered significantly different at a 5% error rate for the overall pattern.

Previously the group means of the logarithms of TCE data were found to be significantly different, based on the overall likelihood-ratio test. Individual comparisons between group means are reported by the tests on regression slope coefficients. To describe which groups differ from others at an overall rate of $\alpha = 0.05$, the p-values for individual comparisons must be less than $0.05/3 = 0.017$. The slope coefficient of 1.40 for MedEq1 (Medium Density) is the difference between the mean of logarithms for the medium-density group and the mean of logarithms for the low-density group (the reference group with values of 0 and 0 for the two binary variables). The mean for the medium density group is 1.40 log units greater than the mean for the low density group. This difference is tested using the Z statistic in Minitab®, equivalent to the partial t-test for the slope in a regression equation. The resulting test statistic in the printout is $Z = 1.25$, with a p-value of 0.210, much larger than 0.017. Therefore the mean logarithm of TCE for the low and medium density groups are not significantly different at an overall alpha of 5%. Similarly, the coefficient of 3.06 for HiEq1 estimates the difference in the mean of logarithms for the High and Low density categories. It is significantly nonzero, with a p-value of 0.007. Therefore a typical ratio is that the mean logarithm (geometric mean) for the high density group is $e^{3.06} = 21.3$ times higher than the geometric mean for the Low density group.

To test the third comparison, the difference between the Medium and High density groups, group assignments need to be recoded to set either the Medium or High groups as the reference category. Then a slope coefficient is computed for the medium versus high group comparison (not shown). The p-value is large, and therefore the difference is not significant. The overall pattern seen by the likelihood ratio test is that TCE concentrations differ between the high and low density areas at a 5% error rate, but neither are significantly different from concentrations in the medium density area.

Bonferroni's procedure for multiple comparisons is a "conservative" method, because the individual $\alpha$-levels may be lower than necessary in order to achieve an overall 5% error rate. An alternative to Bonferroni's multiple comparison procedure is Tukey's honest significant difference test (Zar, 1999). Tukey's test uses the sample sizes in each group to adjust the distances by which two means must be separated in order to call them significantly different. The test statistic q for comparing the means of any two groups (groups a and b) is shown in equation 10.4:

$$q = \frac{\bar{x}_b - \bar{x}_a}{\sqrt{\frac{s^2}{2}\left(\frac{1}{n_a} + \frac{1}{n_b}\right)}} \tag{10.4}$$

where s is the scale coefficient printed in the regression output from maximum likelihood regression. The calculated statistic q is compared to a table of $q_{\alpha,v,k}$, the Studentized range statistic, which is a function of $\alpha$ (the overall error rate), $v$ (the error degrees of freedom $= n - k$), and k (the total number of groups to be compared). However this test is not currently set up to function under MLE software for survival analysis, so it must be computed by hand.

## Nonparametric methods

Several options are available for making nonparametric comparisons between the distribution functions of three or more groups of censored data. These tests do not assume the data follow any particular distribution, and therefore no transformations are required prior to computing the tests. However, the data still need to be manipulated in some way. For contingency tables, the response data are put into two categories, below and above the highest detection limit. For a Kruskal-Wallis test, all data below the highest detection limit are put into the same category while the data above are used as recorded. For the generalized-Wilcoxon score test, the data must be flipped to be right-censored because software for score tests accept only right-censored values. Flipping censored data was previously discussed in Chapter 2.

*Contingency tables*

If the censored response variable is collapsed into two values, below and above the maximum detection limit, contingency tables will test whether the proportion of values in those two categories changes among groups. Collapsing data into two response categories loses information, but (depending on where the highest detection limit is located) the information that remains may be the major component of what is available. Contingency tables are easily understood, and easily illustrated with a simple bar chart. The proportions above and below the highest detection limit are unambiguous, and the test results definitive. The test is available in all commercial statistical software. This is the simplest test to perform with a censored response variable, but it has less power than score tests. The mechanics of the method are found in many statistics textbooks, including Conover (1999).

For the TCE data, observations above the highest detection limit of 5 µg/L are assigned a unique value, either numeric or text, and those below assigned a second value. These values are not ordinal – the test does not require one value to be higher than the other. It only tests whether the percentage of response values in each category is similar across all groups. Figure 10.7 shows the observed percentages of TCE concentrations above 5 µg/L for the three density groups.

A contingency table test of the TCE data can be produced by invoking the Minitab® macro %detects by typing

> %detects c1 c2 c3

where c1 is the column of data/detection limits, c2 is the indicator variable for censoring, and c3 is the group assignment. This produces the output:

```
Pearson Chi-Square = 9.238, DF = 2, P-Value = 0.010
Likelihood Ratio Chi-Square = 11.697, DF = 2, P-Value = 0.003
```

The two chi-square tests use alternate equations for comparing the observed number of counts to the expected number of counts in each cell of the table. Expected counts are the numbers expected when the null hypothesis is true. If the observed counts are similar to expected, the test statistic is small and the null hypothesis is not rejected. If the observed counts differ from what is expected, the test statistic is large and the null hypothesis is rejected. Both tests indicate that the proportions of data above 5 µg/L differ among the three land-use groups, with a p-value of no more than 0.01. The chi-square test detected differences that were obscured when an analysis of variance on substituted data was performed. If a "quick and dirty" test is required, this test of proportions is just as quick, and far less dirty, than substitution.

*Kruskal-Wallis test*

A second approach to analyzing these data without a distributional assumption is the Kruskal-Wallis (KW) test. The KW test determines whether the distribution functions (cdfs) of three or more groups of data are similar, or if at least one is different. The test is applied to censored data by setting all observations below the

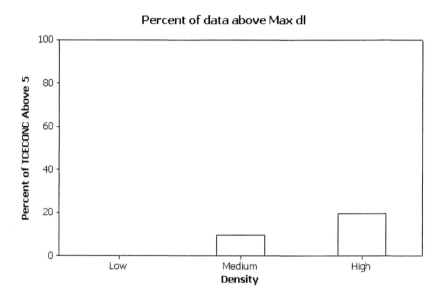

**Figure 10.7.** Percent of TCE concentrations above the highest detection limit of 5 mg/L for three land use categories (data from Eckhardt et al., 1989).

highest detection limit to the same value. When ranked, these observations become tied at the lowest rank. Data above the highest detection limit are ranked using the same method as for uncensored data. If survival analysis software is not available, the KW test provides a valid nonparametric alternative. However, score tests in survival analysis software will provide greater power for multiply-censored data, without re-censoring data to the highest detection limit.

For the TCE data, all observations below the highest detection limit of 5 μg/L are assigned the same value, any value less than 5. All <1, <2, etc., as well as all detected values of 1, 2, 3, and 4 are assigned an identical low number to represent that they are less than 5. The Minitab® macro %censKW assigns a −1 to these lowest values. In this way, all values censored below the highest detection limit are given tied ranks. A negative number is used to reinforce that this is an assigned value, rather than an actual measurement. The KW p-value of 0.01 shows that even with approximately 80% censoring, differences among the cdfs of the three groups can be discerned. As mentioned in Chapter 9, a large proportion of censoring is not in itself a reason to avoid performing a rank-based test like the KW test. If the proportion of censoring differs significantly among the groups, as it does here, that difference can be discerned by the test.

```
Kruskal-Wallis Test on TCEConc-

Density-    N      Median    Ave Rank        Z
Low         25     -1.000    109.0        -1.11
Medium      130    -1.000    120.4        -0.84
High        92     -1.000    133.2         1.56
Overall     247              124.0

H = 2.95   DF = 2    P = 0.228
H = 9.17   DF = 2    P = 0.010   (adjusted for ties)

Use the tie adjustment.   All values below the max dl
were set as tied at -1.
```

Note the estimated median of –1 for all three groups results from the assignment of that value to all nondetects. The medians should be considered to be <5.

*Wilcoxon Score Test*
The Wilcoxon score test for three or more groups is an extension of the score test for two groups presented in Chapter 9. Like the KW test, score tests determine whether distributions (cdfs) of groups are the same, or if at least one is different. Unlike the KW test, score tests extract more information from the data by assigning estimated percentiles (or scores) to detected observations falling between multiple censoring thresholds. When data have multiple detection limits, a score test will have more power (ability to see differences between cdfs) than the KW test because no additional censoring is required. Without censoring the score test would be identical to the Kruskal-Wallis test.

Survival analysis software is programmed to compute score tests only for right-censored data, so environmental data must first be flipped before testing. The resulting p-values will be identical to those that would be computed if left-censored data were allowed. Flipping data for nonparametric tests merely changes the order of ranking from high to low instead of from low to high.

In Minitab®, the TCE data were first flipped by subtracting each concentration from 400 to produce a right-censored variable "FlipTCE". The generalized Wilcoxon test is computed on FlipTCE using the

Reliability/Survival > Distribution Analysis (Right Censoring) > Nonparametric Distribution Analysis

command.

Comparison of Survival Curves

Test Statistics

| Method | Chi-Square | DF | P-Value |
|---|---|---|---|
| Log-Rank | 16.2794 | 2 | 0.000 |
| Wilcoxon | 16.0761 | 2 | 0.000 |

The generalized Wilcoxon test statistic is 16.07, with a corresponding p-value of < 0.001, indicating that the distributions of concentrations differ significantly among the three groups. The lower p-value of the score test as compared to the KW test (p = 0.01) illustrates the greater power of the score test, as it uses the information below the highest detection limit of 5 µg/L more efficiently.

Figure 10.8 presents the survival function plot of the three groups of TCE concentrations. Survival functions plot the cumulative probabilities of exceeding values of right-censored data. These are also the probabilities of being below the left-censored concentration values, and so are the percentiles of TCE. Survival plots are simply plots of the cdf of the original concentration data, plotted right to left instead of left to right, and with the flipping constant (here 400) at a concentration of zero. By drawing a line across the graph at probability = 0.9 (the 90th percentile), the graph shows that the high density group has larger concentrations (smaller flipped values) than the other two groups. This reflects the larger proportion of detected concentrations originally seen in the boxplots of Figure 10.1. The spread between

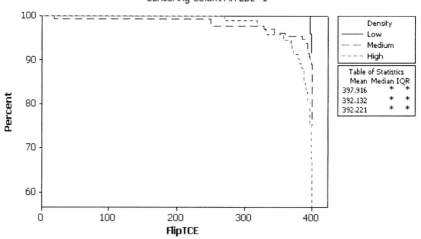

**Figure 10.8.** Survival functions of flipped TCE concentrations for three land use categories. Differences appear substantial between the 99th and 80th percentiles.

the curves of the three survival functions is the significant difference identified by the Wilcoxon score test.

*Multiple comparison tests*

Multiple comparison procedures are a logical next step following rejection of the null hypothesis of similarity by the Wilcoxon score test. To date, the only approach available with commercial software for nonparametric multiple comparisons for censored data is to perform a series of two-group score tests between each pair of groups. If the p-value is less than the Bonferroni individual comparison level obtained using equation 10.3, the two groups can be declared to have different distribution functions at the chosen overall error rate.

For the k=3 groups of TCE data there are g=3 pairwise comparisons. Using Bonferroni's method, each pairwise comparison must have a p-value below 0.05/g = 0.017 for the group cdfs to be considered different at the overall error rate of 5%. Computing each of these pairwise comparisons for the flipped TCE data using the score test procedure in Minitab® results in:

Low versus Medium

| Method | Chi-Square | DF | P-Value |
|--------|-----------|----|---------|
| Wilcoxon | 0.68890 | 1 | 0.407 |

Low versus High

| Method | Chi-Square | DF | P-Value |
|--------|-----------|----|---------|
| Wilcoxon | 7.09906 | 1 | 0.008 |

Medium versus High

| Method | Chi-Square | DF | P-Value |
|--------|-----------|----|---------|
| Wilcoxon | 11.5275 | 1 | 0.001 |

Both the low and medium density areas have significantly different TCE concentrations than the high density area, using nonparametric score tests. The low and medium density areas are not different from each other.

Other methods for nonparametric multiple comparisons exist, but are not coded into survival analysis software. These include the Dwass et al. test (Hollander and Wolfe, 1999, p. 240), a test computed using ranks of the data. For censored data the Wilcoxon scores can be used rather than the ranks. A second nonparametric multiple comparison test is the slippage test (Conover, 1968), a test that counts the number of observations in a group that are greater than the highest observation in the next lowest group. If more observations than expected exceed the top of the next group, that group has 'slipped' significantly lower and a difference is declared. Because the test statistic is based primarily on observations at the high end of each group, left censoring is not often an issue when computing the slippage test.

To illustrate the slippage test, the k=3 residential-density groups of TCE concentrations are ordered based on the magnitude of their maximum observations (see Table 10.2). By this criteria the Medium-density group is 'higher' than the High-density group. Starting with the highest group, the number of observations $r_i$ is

**Table 10.2.** Computation of the slippage test for the TCE data.

| Density | Medium | High | Low |
|---|---|---|---|
| maximum TCE | 382 | 130 | <5 or 4 |
| # exceedances $r_i$ | 3 | 18 | – |
| sample size n | 130 | 92 | 25 |
| Prob (cdf$_i$ = cdf$_j$) | 0.275 | 0.016 | – |

counted that exceed the maximum of the next lowest group. For example, three TCE concentrations in the Medium-density group exceed 130 µg/L, the highest concentration in the High density group.

The probability that the distribution of the ith group $Gp_i$ is identical to the distribution of the next lowest group $Gp_{i+1}$, is (Conover, 1968):

$$\text{Prob}(Gp_i > Gp_{i+1}) = \left[ \frac{n_i - 1}{\left(\sum_{j=i}^{k} n_j\right) - 1} \right]^{r-1} \tag{10.5}$$

This is equivalent to the p-value for the individual comparison test.

For the comparison between the (i=1) Medium-density group and the next lowest (High-density) group,

$$\text{Prob}(Gp1 > Gp2) = \left[ \frac{129}{(130 + 92 + 25) - 1} \right]^{2} = 0.275$$

and between the (i=2) High-density group and the (i=3) Low-density group,

$$\text{Prob}(Gp2 > Gp3) = \left[ \frac{91}{(92 + 25) - 1} \right]^{17} = 0.016$$

The p-values get smaller as the test statistic r, the number of exceedances, gets larger. When r equals 1, the p-value resulting from equation 10.5 equals 1.

Individual comparisons are declared different if their p-values are less than $\frac{\alpha}{(k-1)}$, where k is the number of groups and a is the desired overall error rate. For a=0.05 and k=3, the p-values are compared to an individual error rate of $\frac{0.05}{2}$, or 0.025. Therefore, the slippage test concludes that there is a significantdifference between the High and Low density groups, but not between the Medium and High density groups.

The results of different multiple comparison tests do not always agree. The slippage test and individual Wilcoxon tests did not agree as to which groups differed from others. The slippage test focuses more on the high end of each group than does the Wilcoxon test, and is insensitive to differences occurring in the central and lower portions of each group. The results of the slippage test seem to agree with the visual impression of the boxplots of Figure 10.1, which only show the high-end values above the highest detection limit. The choice of which test to use should be based on which characteristic of each group is the most appropriate to distinguish.

Perhaps in future releases, software will provide more in the way of nonparametric multiple comparison procedures for censored data.

## Summary

Tests for differences in the distributions of three or more groups of censored environmental data can be carried out in several ways. These methods are direct extensions of the two-group tests of the previous chapter. Substitution methods perform poorly, resulting in both false rejections and false non-rejections of the null hypothesis. For multiply-censored data in particular, substitution of values linked to the magnitude of the detection limit should be completely avoided.

Maximum likelihood estimation provides parametric tests of hypotheses without substitution. When the assumption of a distributional shape such as the lognormal appears reasonable, MLE methods perform well. These MLE analogues of analysis of variance are conducted on left-censored data using methods for "arbitrary" or interval-censored data, where nondetects are considered to be within the interval 0 to the detection limit. Group differences can be tested using either a "factor" designation for the group variable, or with binary variables representing group membership. Tests for whether regression slopes are different from zero are in this case parametric tests for differences in group means of censored data.

Nonparametric score tests extract the maximum information for determining group differences from multiply-censored data without assuming a distributional shape. The generalized Wilcoxon test, an extension of the Kruskal-Wallis test to multiply-censored data, efficiently detects differences in the distributions of groups without assuming normality. Currently, data must be flipped into a right-censored format prior to testing, due to software limitations. When software for score tests is not available, a Kruskal-Wallis test may provide sufficient power to distinguish between groups, but data below the highest detection limit must be re-coded to a single value. An even simpler method is to represent all data as binary values, either below or above the highest detection limit, and perform a contingency table analysis. This will provide a more dependable answer than using substitution, without dangerous assumptions.

Reasonable methods for plotting multiply-censored data are available, starting with survival function plots, which are actually plots of the cdfs of the censored data. These plots show the results of the Wilcoxon score test, much as censored boxplots do for the Kruskal-Wallis test and bar graphs of percentages above the highest detection limit do for contingency table analyses.

# Exercises

**10-1**    Golden et al. (2003) measured concentrations of lead in the blood and in several organs of herons in Virginia, in order to relate those concentrations to levels found in feathers. The objective was to determine whether feathers were a sensitive indicator of exposure to lead. If so, feathers could be collected in the future so that the birds would not need to be sacrificed in order for their exposure to lead to be evaluated. The herons received different doses of lead. Exposure was categorized into one of four groups: a control group receiving no additional lead, and groups receiving 0.01, 0.05, and 0.25 mg lead per g of body weight. Determine whether the lead found in feathers of these birds differed among the four exposure groups at an $\alpha$ = 0.05 level. If so, run a multiple comparison test to determine which groups differ from the others. Use the methods of this chapter (not substitution!) to test for differences, using the data found in Golden.xls.

**10-2**    Brumbaugh et al. (2001) measured mercury concentrations in fish of approximately the same trophic level across the United States, as well as characteristics for the watersheds they lived in. The data are found in HgFish.xls. The variable "LandUse" reflects the dominant land-use within the watershed, and includes categories of Ag/Forested (or "A/F"), Ag, Mining, Urban, and Background. Test to see if mercury concentrations in fish (variable "Hg") differ among the 5 land-use categories. The mercury concentrations have been censored at three detection limits, as indicated with a value of 1 for the variable "HgBDL1". Note that the A/F land-use includes watersheds containing the largest proportion of wetlands.

**10-3**    Yamaguchi et al. (2003) measured concentrations of PCBs in fish collected at four sites draining to the Thames River, UK. Three sites are below Swindon, an active industrial area draining to the Ray River, a tributary to the Thames. The fourth site, Burford, is on the Windrush tributary and not downstream of Swindon. Test whether PCB concentrations are the same or different in fish at the four sites, using both parametric (censored regression) and nonparametric (Wilcoxon score) tests. The data are found in Thames.xls.

# 11

# CORRELATION

How strong is the association between two variables? As X increases, how likely is it that Y consistently increases, or decreases? The strength of the association or dependence between two variables, how predictably they co-vary, is measured by a correlation coefficient. Correlation coefficients vary between values of −1 and +1, and centered at zero, the value where no correlation is observed. As the coefficient moves away from 0 in either direction, evidence for correlation increases. At values of +1 or −1 perfect dependence is observed, with the sign denoting whether the two variables move in the same direction (+) or in opposite directions (−). One of the most common uses for correlation coefficients in environmental studies has been for investigation of trends. When X represents time, the test for significance of the correlation between Y and X is a test for a trend in Y. This is the context for the example data set used in this chapter.

Consider the following values of dissolved iron concentrations presented in Table 5 of Hughes and Millard (1988), collected during summers from the Brazos River, Texas:

| Dissolved Iron (Y): | 20 | <10 | <10 | <10 | <10 | 7 | 3 | <3 | <3 |
|---|---|---|---|---|---|---|---|---|---|
| Time, in years (X): | 1977 | 1978 | 1979 | 1980 | 1981 | 1982 | 1983 | 1984 | 1985 |

Do summer dissolved iron concentrations exhibit a trend during this period?

**Types of correlation coefficients**

The traditional (parametric) correlation coefficient is Pearson's r. Pearson's coefficient measures the linear correlation between Y and X. As seen in equation 11.1, Pearson's r involves computing both the mean and standard deviation for both X and Y, as well as a measure of the distance each observation is from its mean. All three items are difficult to calculate with censored data. Software for censored data does not attempt to compute Pearson's r.

$$\text{Pearson's r} = \frac{1}{n-1}\sum_{i=1}^{n}\left(\frac{x_i - \bar{x}}{s_x}\right)\left(\frac{y_i - \bar{y}}{s_y}\right) \qquad (11.1)$$

Spearman's rho is a nonparametric correlation coefficient that is computed by calculating Pearson's r on the ranks of the original data. Where those ranks can be computed unequivocally, as when there is only one detection limit, Spearman's rho provides a feasible alternative to Pearson's r for censored data. For multiple detection limits, however, computation of Spearman's rho involves similar difficulties to Pearson's r. It is rare that software for censored data attempts to compute Spearman's rho.

Kendall's tau is a second nonparametric correlation coefficient that is commonly used in tests for trend, and can easily be adapted for censored data. No computation of means, or distances from means, are required. Tau is computed for a data set of n (X,Y) pairs as the number of concordant pairs of data ($N_c$) minus the number of discordant pairs ($N_d$), divided by the number of total pairs, or

$$\text{Kendall's } \tau = \frac{N_c - N_d}{n(n-1)/2} \qquad (11.2)$$

Tau is most easily computed by first ranking the data in order of increasing X. Then concordant pairs are those pairs of observations where Y increases as X increases, or where there is a positive slope. Discordant pairs are those where X and Y are going in opposite directions, or where there is a negative slope. Pairs which are tied are assigned a 0. When many ties occur, Kendall (1955) proposed adjusting the denominator for the number of tied observations. This is called Kendall's tau-b. Tau-b is the form most often used, and most applicable to, correlation of censored data. Tau-b is discussed further in the section on nonparametric methods.

Finally, there are correlation coefficients that can be computed for binomial data, data whose values are categorized as a 0 or 1. These coefficients are applied to data analyzed by contingency tables (see Chapter 10). For censored data, if all values below a single detection limit are assigned a 0, and all detected observations assigned a 1, these coefficients indicate the correlation of detections to a grouping variable. Does the frequency of detection change from one group to the next? Two coefficients useful in this situation are again Kendall's tau-b, and a measure called the Phi coefficient.

In parallel to the discussions in other chapters, methods for computing correlation coefficients using substitution, maximum likelihood, and nonparametric methods are now illustrated.

## Substitution methods

Published papers have reported correlation coefficients, usually Pearson's r, calculated after fabricating values for nondetects. One-half the detection limit is the value most often used. For the Brazos River iron data where a trend in concentration is investigated, substituting the detection limits results in a Pearson's r of –0.89. The hypothesis test (a t-test) for determining whether r = 0 has a p-value of 0.001, indicating a significant downtrend in iron concentrations. When zeros are substituted for all nondetects, r equals –0.46 with a p-value of 0.216, indicating that a trend (correlation) is unlikely. Which of these conclusions is correct? Once again, substitution produces neither accurate nor unique results. There are better ways.

Previous attempts to find a better method for censored correlation include that of Sanford et al. (1993), who used Cohen's maximum likelihood method to estimate values prior to computing correlation coefficients. They found that substituting Cohen's MLE estimate of the mean of censored values for each nondetect worked far better than substitution of a single value such as 0.75 times the detection limit, their discipline's standard procedure. It also worked far better than deleting censored values. Correlation coefficients can be computed directly using MLE, but their paper was an early attempt (in the environmental sciences literature) to deal with the problem of correlation of censored data.

## Maximum likelihood estimation

Users of regression software consider the coefficient of determination ($r^2$) a measure of the quality of the regression equation. Modern textbooks on statistics caution against over-dependence on this statistic, noting that poor regression models can have high $r^2$, and vice-versa. However, it is a reasonable measure if the data follow a straight line pattern, the residuals are close to a normal distribution, and no single point overly affects the position of the regression line (Ryan, 1997). Maximum likelihood methods are capable of producing several measures similar in concept to $r^2$, the most popular being the likelihood $r^2$. Computation of the likelihood $r^2$ (called the generalized $r^2$ by Allison, 1995) is very different from the process used by standard least-squares regression with uncensored data. The log-likelihood statistic is determined for the regression of Y versus one or more X variables (the "full model") computed using maximum likelihood. A second log-likelihood statistic is determined for the "null model", using no X variables. For the null model, the variation of Y is only the noise around the mean of Y. The difference between these two likelihood statistics is multiplied by -2 to produce the "-2 log likelihood" or $G_0^2$. $G_0^2$ measures the increase in the likelihood of producing the observed pattern of Y when the relationship with the X variables is taken into consideration. For large values of $G_0^2$ the null hypothesis of no linear relationship between Y and the X variables is rejected, and the slope of the relationship is determined to be nonzero. In order to determine its significance, $G_0^2$ is compared to a chi-square distribution with degrees of freedom equal to the number of X variables in the full model. For correlation, only one X variable is usually considered at a time.

The value of $G_0^2$ is used to compute the likelihood $r^2$:

$$\text{likelihood } r^2 = 1 - \exp\left(-\frac{G_0^2}{n}\right) \qquad (11.3)$$

where n is the number of (x,y) paired observations. For large values of $G_0^2$ the value of $r^2$ will be close to 1. The likelihood r correlation coefficient is the square root of the likelihood $r^2$, with algebraic sign identical to that for the slope of the relationship.

$$\text{likelihood } r = \text{sgn}(slope) \bullet \sqrt{1 - \exp\left(-\frac{G_0^2}{n}\right)} \qquad (11.4)$$

This correlation coefficient best expresses the strength of the relationship between Y and X as measured by maximum likelihood. For the dissolved iron data, maximum likelihood was performed assuming a linear relationship between concentration and time, and assuming that residuals followed a normal distribution. For this small data set, those assumptions are difficult to verify. The resulting $G_0{}^2 =$ 7.35, the slope was negative, and from equation 11.3, the likelihood $r^2 = 0.56$. The likelihood r is therefore $= -0.75$, and the test of the null hypothesis that r = 0 is identical to the test for whether the slope equals 0, just as in ordinary linear regression. Performing regression by maximum likelihood (for further information, see Chapter 12), the slope $= -1.73$ and its p-value $= < 0.001$. Therefore we reject the null hypothesis that X and Y are uncorrelated, and conclude that there is a significant linear correlation.

For this example, the value of the likelihood r ($-0.75$) happens to be between the values for Pearson's r computed by substitution of the maximum (detection limit) and minimum (zero) values possible for censored data. This may not always the case, due to the vagaries of substitution methods' computation of standard deviations. The likelihood correlation and associated significance test have the advantage that they can be computed for data subject to multiple censoring limits. Their validity will depend on whether there are enough data to determine that the relationship is in fact linear, and whether the variation around the line is normally distributed. These are the same constraints required of Pearson's r.

## Nonparametric methods

### Spearman's rho

Spearman's rho is a rank-transform measure of the monotonic association between Y and X (Helsel and Hirsch, 2002). A positive monotonic correlation means that as X increases, Y consistently increases, although the pattern may or may not be linear. Thus rho (and tau) are more general measures of correlation than is Pearson's r, which measures the strength of linear associations.

Rho is calculated by ranking each variable separately, and then computing Pearson's correlation coefficient on the ranks. The ranks when data have only one detection limit are unique and can easily be used to compute rho, but direct computation of rho for multiply-censored data is not possible. To illustrate the computation of rho, ten paired observations from the atrazine data of Junk et al. (1980) are listed in Table 11.1. Their ranks are computed and reported in the two right-hand columns. Values less than the detection limit are considered ties and given their average rank; the four June observations at <0.01 are each given the rank of 2.5, the average of ranks 1 through 4.

Computing Pearson's r on the two columns of ranks produces a Spearman's rho of 0.74 for the atrazine data, as shown in the following output from Minitab®:

```
Pearson correlation of rankJune and rankSept = 0.743
P-Value = 0.014
```

The small p value (less than 0.05) and positive slope are evidence that concen-

**Table 11.1.** A subset of the atrazine concentrations reported by Junk et al. (1980). Used as an example of data having one detection limit.

| June | September | Rank of June | Rank of September |
|------|-----------|--------------|-------------------|
| 0.38 | 2.66 | 10 | 10 |
| 0.04 | 0.63 | 8 | 8 |
| < 0.01 | 0.59 | 2.5 | 7 |
| 0.03 | 0.05 | 6.5 | 5 |
| 0.03 | 0.84 | 6.5 | 9 |
| 0.05 | 0.58 | 9 | 6 |
| 0.02 | 0.02 | 5 | 4 |
| < 0.01 | 0.01 | 2.5 | 3 |
| < 0.01 | < 0.01 | 2.5 | 1.5 |
| < 0.01 | < 0.01 | 2.5 | 1.5 |

trations in June are correlated with concentrations in September. Wells with high concentrations in June are generally also high in September. This correlation may or may not be linear, but it is at least monotonic.

When Y and X are both censored and have one detection limit each (though not necessarily the same detection limit), rho can easily be computed. This was the case with the atrazine data. When X or Y have multiple detection limits, however, the data must be re-censored at the highest detection limit in order to compute Spearman's rho. This is because it is difficult to judge the relative rank of a <10 as opposed to a 5, or a <1. Though some adaptations of rho have been suggested in this situation, such as using Kaplan-Meier scores in the place of ranks, no standard software implements adaptations of rho, and its characteristics are not well described in the literature. However there is a nonparametric correlation coefficient for the case of multiple detection limits — Kendall's tau.

*Kendall's tau*

Kendall's tau has a distinct advantage over Spearman's rho — it can easily be computed for multiply-censored data. Tau is computed by comparing all pairs of observations, counting the number of positive slopes minus the number of negative slopes, and dividing by the total number of pairs of observations (equation 11.2). If all slopes are positive, tau equals 1.0. If all slopes are negative, tau equals –1.0. When they are half and half, as expected on average when X and Y are unrelated, tau equals 0. Kendall (1955) gives a complete description of the coefficient, including a modification when tied values occur that is called tau-b. Brown, Hollander and Korwar (1974) adapted tau for use with censored data in heart transplant studies. To compute tau-b (equation 11.5), concordant pairs $N_c$ and discordant pairs $N_d$ are computed for all pairs whose differences are clear. For example, after first sorting the data by X, a change in Y from <1 to 5 is a clear increase, a concordant pair. Pairs where Y does not change, X does not change, or where the change is indeterminate, such as <1 to <10, are considered ties. Ties do not contribute to the numerator of tau-b, and are subtracted from the number of possible pairwise comparisons in the denominator of tau-b.

$$\text{Kendall's } \tau_b = \frac{N_c - N_d}{\sqrt{\left(\frac{N(N-1)}{2} - \#\text{ties}_x\right)\left(\frac{N(N-1)}{2} - \#\text{ties}_y\right)}} \quad (11.5)$$

where $\#\text{ties}_x$ is the number of ties in the X variable and $\#\text{ties}_y$ is the number of ties in the Y variable. If there are no ties in X or Y, equation 11.5 simplifies to equation 11.2.

Consider again the multiply-censored dissolved iron concentrations of Hughes and Millard (1988), ordered by increasing values of X. Is there a correlation, a trend over time, in these data?

| Dissolved Iron (Y): | 20 | <10 | <10 | <10 | <10 | 7 | 3 | <3 | <3 |
|---|---|---|---|---|---|---|---|---|---|
| Time, in years (X): | 1977 | 1978 | 1979 | 1980 | 1981 | 1982 | 1983 | 1984 | 1985 |

To compute Kendall's tau, the first Y observation is compared to all subsequent observations. Concordant observations are those where Y increases (a positive slope, as X is increasing). Assign a + to those comparisons. Discordant observations are those where Y decreases as X increases, a negative slope. Assign a − to those observations. Comparisons to the first observation of Y = 20 are shown below their respective observation. All are decreasing.

| Dissolved Iron (Y): | 20 | <10 | <10 | <10 | <10 | 7 | 3 | <3 | <3 |
|---|---|---|---|---|---|---|---|---|---|
| sign of difference: | | − | − | − | − | − | − | − | − |

Multiple detection limits have posed no problem for these comparisons. Going from a 20 down to a <10 is a decrease, as is going down to a <3.

Next, the second Y observation is compared to all subsequent values. Here none of the comparisons is clearly an increase or decrease. It is impossible to determine whether a <10 is higher or lower than a 7, or a <3. All of the comparisons with the second observation are given zeros, neither concordant nor discordant.

| Dissolved Iron (Y): | 20 | <10 | <10 | <10 | <10 | 7 | 3 | <3 | <3 |
|---|---|---|---|---|---|---|---|---|---|
| sign of difference: | | − | − | − | − | − | − | − | − |
| | | | 0 | 0 | 0 | 0 | 0 | 0 | 0 |

The scores for all subsequent comparisons are shown below:

| Dissolved Iron (Y): | 20 | <10 | <10 | <10 | <10 | 7 | 3 | <3 | <3 |
|---|---|---|---|---|---|---|---|---|---|
| sign of difference: | | − | − | − | − | − | − | − | − |
| | | | 0 | 0 | 0 | 0 | 0 | 0 | 0 |
| | | | | 0 | 0 | 0 | 0 | 0 | 0 |
| | | | | 0 | 0 | 0 | 0 | 0 | 0 |
| | | | | 0 | 0 | 0 | 0 | 0 | 0 |
| | | | | | − | − | − | − |
| | | | | | | − | − | − |
| | | | | | | | | 0 |

There were 0 concordant pairs and 13 discordant pairs, so the numerator for tau-b = −13. There were no ties among the values of X, but 23 ties in the comparisons between Y observations, including the comparisons that were unclear due to censoring. Kendall's tau-b for these data is therefore

$$\text{Kendall's} = \tau_b = \frac{0-13}{\sqrt{\left(\frac{9(8)}{2}-0\right)\left(\frac{9(8)}{2}-23\right)}} \quad -0.60$$

The test for significance of Kendall's tau is computed using $S = N_c - N_d$ as the numerator of the test statistic, and the standard error of S as the denominator. The square of the standard error is the variance of S, which for the case of no ties equals

$$\text{var}[S] = \frac{N(N-1)(2N+5)}{18}$$

The test statistic Z is compared to values of the standard normal distribution.

$$Z = \frac{S - sgn(S)}{\sqrt{\frac{N(N-1)(2N+5)}{18}}} \tag{11.6}$$

where sgn is the algebraic sign function. The numerator uses a continuity correction of sgn(S) to better calculate p-values from a smooth normal distribution function, adding or subtracting a value of 1 from S (Kendall, 1955).

With many ties resulting from comparisons among censored values, a correction is required for the variance of S. Software which performs this adjustment is crucial for censored data, because nondetects produce many tied comparisons. For censored data, the tie correction is more complex than if all ties resulted from detected observations. This is because when there are 4 observations tied at 10, all comparisons between the 4 observations are ties, so there are $4(3)/2 = 6$ ties among the 4 observations. Tie corrections in commercial software usually compute the number of ties as $\#obs \bullet (\#obs - 1)/2$ for a set of tied observations. But there are fewer ties among a set of 4 observations which include both censored and uncensored values, such as <1  4  7 <10. In this case there are 3 ties: [<1, <10], [4, <10], and [7, <10]. The other 3 comparisons are known to be pluses: [<1, 4], [<1, 7], and [4, 7].

The ckend macro for Minitab® computes Kendall's tau-b and its significance test as described above, adjusting both tau and Z for ties originating from multiply-censored data. The output from the ckend macro for the dissolved iron data of Millard and Hughes (1988) is:

```
S            -13.0000
tau          -0.361111
taub         -0.600925
z            -1.50787
pval          0.131587
```

For these data, tau-b is not significantly different from zero (p = 0.13). There is not sufficient evidence of monotonic correlation to declare that there is a trend in concentration.

For censoring at a single detection limit, Kendall's tau is easy to compute. For the subset of atrazine observations of Table 11.1, all values at <0.01 are the lowest observations with the lowest ranks, as shown in the right hand columns of the table. As with all nonparametric correlation coefficients, ranks are computed separately for each variable. Within each variable, ranks of censored observations are tied. The resulting tau-b is 0.636 with a two-sided p-value of 0.018, There is a significant association between June and September atrazine concentrations in this subset as determined by Kendall's tau.

```
S        26.0000
tau       0.577778
taub      0.635851
z         2.36651
pval      0.0179568
```

*The Phi coefficient*

For singly-censored data, or data re-censored to the highest detection limit, values can be classified into two categories, either above or below the detection limit. A variety of methods for binomial data, represented perhaps as 0s and 1s, are available to the data analyst. The phi coefficient is a correlation coefficient for paired binomial data – in fact it is identical to Pearson's r computed on data represented as one number per class (Conover, 1999, p. 234). Methods for binomial data are most useful when values are severely censored, with more than about 80% nondetects. In this situation the preponderance of information contained in the data is represented by the proportions in each category, rather than by numerical values of individual observations. Binomial methods efficiently capture this proportion information. Phi may be the easiest measure of association to explain when data are severely censored, though it is not required in that situation – rho or tau may also be computed for severely censored data. For example, Kolpin et al. (2002b) computed rho for the relation between detected/nondetected observations and values of single explanatory variables.

To compute phi, consider variables X and Y, each consisting of observations classed as either High or Low. Four combinations of the two variables are possible, as shown below.

| X | Y |
|---|---|
| High | High |
| High | Low |
| Low | High |
| Low | Low |

Phi will have the largest positive value when both variables are High together, and both Low together. Large negative values result from consistent classifications into opposite categories. A mix of conditions will produce phi values close to 0, as with any correlation coefficient. Counts of occurrences of the four combinations may be visualized as a table (Table 11.2). The order of the columns and rows of the table should be such that cells a and d represent positive correlation, and cells b and c negative correlation.

**Table 11.2.** Two variables classified into two categories each. A 2×2 contingency table.

|  |  | (June) |  |
|---|---|---|---|
|  |  | Low | High |
|  | Low | a = 2 | b = 0 |
| (September) | High | c = 2 | d = 6 |

Table 11.2 contains the counts of pairs of low (<0.01) and high (detected) values for the 10 observations in the atrazine data of Table 11.1. There were 6 pairs where atrazine was detected in both June and September, and 2 where both had nondetects. There were two pairs where atrazine was below the detection limit in June, but above in September (even though one was just barely above, detected at 0.01). There were no cases where atrazine was detected in June but not in September.

If $r_1$ = the sum of counts in row 1, $r_2$ the sum of counts for row 2, $c_1$ the sum of counts for column 1 and $c_2$ the sum of counts for column 2, then the phi coefficient is computed as

$$\phi = \frac{ad - bc}{\sqrt{r_1 r_2 c_1 c_2}} \tag{11.7}$$

For the Table 11.1 data $\phi = \dfrac{2 \bullet 6 - 0 \bullet 2}{\sqrt{2 \bullet 8 \bullet 4 \bullet 6}} = \dfrac{12}{\sqrt{384}} = 0.61$.

The test statistic for whether phi is significantly different from 0 is computed by multiplying phi by the square root of n, where n is the number of paired observations, and comparing the product to a standard normal distribution (Conover, 1999). If we are looking for correlation in only one direction, a one-sided test is performed. When both positive or negative correlation would be of interest, a two-sided test is performed. For the Table 11.2 data, the test statistic is. $\phi\sqrt{N} = 0.61\sqrt{10} = 1.93$. For the case where only a positive correlation between June and September concentrations is of interest, the one-sided p = 0.027 and we conclude that there is an association, a positive correlation. The more general two-sided p-value equals twice this, or p = 0.054, right on the edge of the default 0.05 criteria for significance.

The p-value for phi is somewhat larger than (less significant than) for rho and tau. This is because less information is used, and less required, to compute phi. In particular, the rank ordering of values above the detection limit is ignored by phi. All values above the detection limit are only considered equivalent; they are in the same category. Phi is appropriate for data below and above a single threshold for each variable. For multiple detection limits, either use counts above and below the highest detection limit, or instead compute Kendall's tau. There is little reason to prefer phi over tau, and tau will take advantage of the information in the ranks above the detection limits. Of course a threshold higher than the (highest) detection limit, such as at a human or ecologic health criteria, could be used instead to compute phi. But this book is about handling data in reference to detection limits.

### Summary. A comparison among methods

For data with one detection limit per variable, each of the methods in this chapter can be used to estimate the magnitude and significance of bivariate correlation. However, maximum likelihood is limited to data where only the y variable is censored. In order to compute a correlation coefficient by maximum likelihood where "double censoring" of both x and y occurs, censoring in the x variable must be ignored. This can lead to errors. Double censoring can be accounted for directly with Kendall's tau and the phi coefficient, and can be accounted for with Spearman's rho for one detection limit. As the least censored variable for the data in Table 11.1, the Sept concentrations were used as the x variable in MLE regression with the detection limit values used for the two censored observations. Four methods for computing correlation coefficients produce the results in Table 11.3:

**Table 11.3.** Correlation coefficients for the Table 11.1 atrazine data.

|  | Correlation Coefficient | Two-sided p-value |
|---|---|---|
| Likelihood r | 0.99 | 0.000 |
| Spearman's rho | 0.74 | 0.014 |
| Kendall's tau-b | 0.64 | 0.018 |
| Phi | 0.61 | 0.054 |

The likelihood r coefficient uses information in the assumptions that x and y are linearly related, and that the distribution of data around that line follow a normal distribution. These are strong assumptions that are difficult to test when the data set is small or severely censored. One outlier can also badly inflate the likelihood r correlation coefficient, just as it can with Pearson's r for uncensored data, when one observation is a highly influential point. The plot in Figure 11.1 shows that one outlier has likely inflated the likelihood r coefficient to be near 1. The best use of the likelihood r coefficient is for larger data sets that can be evaluated on their adherence to the normality and linearity assumptions, and to the situation of no influential observations. The Table 11.1 data don't fit these assumptions well, and so one of the nonparametric coefficients is a better reflection of the association between the two variables.

Spearman's rho and Kendall's tau are both measures of monotonic correlation, and give similar results. Note the similarity of their p-values. The two coefficients are measured on different scales, with rho on the same scale as the traditional

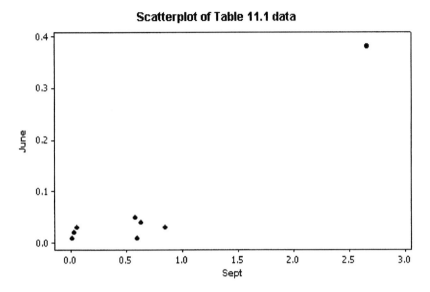

**Figure 11.1.** Scatterplot of the censored data of Table 11.1. Two censored values for Sept. are plotted at their detection limits.

Pearson's r coefficient. Tau is expected to be smaller than rho for the same strength of correlation (Helsel and Hirsch, 2002). Kendall's tau might be preferred due to its ability to be used for multiply-censored data.

Phi measures the association between variables after reducing the data into two categories. This is a heavy price to pay in information content unless the data are strongly censored. The loss of information from classifying all detected observations in this data set as merely "above the detection limit" produces the higher p-value for phi as compared to tau or rho.

For data with multiple detection limits, options are somewhat more limited. Spearman's rho and phi cannot be computed unless the data are re-censored at the highest detection limit, which greatly reduces the information content of the data. If only one variable is censored the likelihood r can be computed by MLE. Kendall's tau handles data even if doubly censored or multiply censored. For the multiply-censored dissolved iron data the likelihood r and Kendall's tau values are listed in Table 11.4:

**Table 11.4.** Correlation coefficients for the summer dissolved iron data.

|  | Correlation Coefficient | Two-sided p-value |
|---|---|---|
| Likelihood r | −0.75 | < 0.001 |
| Kendall's tau-b | −0.60 | 0.132 |

The difference in p-values between the two coefficients is due to the "information" contained in the assumptions of linearity and normality used by the likelihood r coefficient. If the data are assumed to be linear, the four <10s at the early part of the record must be assumed to be at the higher end of their range in order to produce a linear trend with low values at the end of the record. If so, a strong correlation results. This is the linear pattern assumed to occur by the likelihood r coefficient. Kendall's tau makes no assumption of linearity, and allows the uncertainty of the positions of these four values to remain. It therefore reflects more accurately what is known (or more importantly, not known) about the data. The likelihood r coefficient should be used only if the assumption of linearity can be assumed to be true. Based only on the information in this small data set, there is insufficient information to make that assumption; the likelihood r coefficient gives a misleading picture of what the observed data are actually able to tell us.

## For further study

Hughes and Millard (1988) computed the significance test for Kendall's tau in a new and different way than described here, determining all possible permutations of ranks allowed by the censoring scheme and computing tau and its significance test using the average of the possible ranks for each observation. Computations for this "expected rank statistic" method are quite complex, as thousands of permutations are possible for even a moderate sized data set. No commercial software performs this procedure. The values for S and tau will be the same as for the procedures to compute Kendall's tau outlined in this chapter, because tau is defined as the average of possible values for all permutations (Kendall, 1955). The test of significance for the permutation procedure will differ, however, from the Brown-Hollander-Korwar tie correction method used by the Ckend macro. For the summer iron data, Hughes and Millard report a test statistic Z of –2.27 with the corresponding p-value of 0.012, declaring a trend for the period. This is greater evidence for a trend (smaller p-value) than found by the Brown et al. procedure. They state that their expected rank method recovers some of the information lost by the Brown et al. procedure when comparing a <DL to a small detected value or to another censored value. Further investigation of the two methods and their applications to environmental data is warranted.

Oakes (1982) presents the Brown et al. procedure in more detail and applies it to two example data sets. Oakes' results differ from those of the Ckend macro in that he did not use a continuity correction, and did not account for ties in both X and Y. When ties in X occurred, as happened in the example data he presents, he randomly chose an ordering for the tied observations, giving them untied ranks. Therefore his results are not as accurate as those presented here using the Ckend macro, which incorporates ties in both X and Y.

Isobe et al. (1986) first applied the Brown et al. procedure for censored correlation to left-censored (nondetects) data, using examples in the field of astronomy. They argued for routine adoption of these methods in the field of astronomy, much as this book does for the field of environmental sciences.

Akritas and Siebert (1996) derived a test for partial correlation using a partial Kendall's tau. This allows the correlation of X and Y (when one or both include censored values) to be adjusted to account for the influence of one or more covariates. Application is again given to the field of astronomy but the method could be directly applied to censored environmental data.

# Exercises

**11-1**     Golden et al. (2003) measured concentrations of lead in the blood and in several organs of herons in Virginia, in order to relate those concentrations to levels found in feathers. The objective was to determine whether feathers were a sensitive indicator of exposure to lead. If so, feathers could be collected in the future so that the birds would not need to be sacrificed in order for their exposure to lead to be evaluated. Compute a correlation coefficient to determine whether lead concentrations in feathers are associated with concentrations in blood. Note that both have censored values. The data are found in Golden.xls.

**11-2**     Brumbaugh et al. (2001) measured mercury concentrations in fish of approximately the same age and trophic level across the United States. Even so, the size of fish varied due to differences in age and species. Determine whether there is a significant correlation between mercury concentrations ("Hg") and fish length. The data are found in HgFish.xls.

**11-3**     Yamaguchi et al. (2003) measured concentrations of dieldrin and lindane in fish collected at four sites draining to the Thames River, UK. Determine whether concentrations of the two contaminants in fish are correlated. Note that both concentrations contain censored values. The data are found in Thames.xls.

# 12

# REGRESSION AND TRENDS

One of the most frequently used techniques in statistics is linear regression — relating a response variable to one or more explanatory variables by use of a linear model. Estimates of slopes and intercept are computed by least squares, with partial t-tests determining if slope estimates differ significantly from zero. Those that do are worth including in the model. The least-squares procedure produces optimal estimates of slope and intercept if the residuals, the distances in the y-direction between observations and the fitted line, approximately follow a normal distribution, have constant variance across the range of x values, and are linear. The resulting regression line is a conditional mean of y given x; the parametric regression line can be thought of as a "linear mean". If one of the explanatory variables is a measure of time, the test for significance of the slope of that variable is a test for (temporal) trend.

A nonparametric analogue to linear regression, commonly used in trend analysis of environmental data, is the Theil-Sen line and slope estimator (Helsel and Hirsch, 2002). The slope and intercept for the Thiel-Sen line are estimated by an entirely different method than least squares. The slope is related to the nonparametric Kendall's tau correlation coefficient – the slope is the ratio $\Delta y / \Delta x$ that, if subtracted from the response variable y, would produce an altered data set having a Kendall's tau correlation coefficient of zero. The Theil-Sen slope is usually computed as the median of all possible pairwise slopes between observations, and is significantly different from zero when the Kendall's tau correlation coefficient is significantly different from zero. The Theil-Sen line is a "linear median" not strongly influenced by the presence of outliers. When the explanatory variable is a measure of time, the significance test for the slope is a test for (temporal) trend.

There are several valid methods for incorporating nondetects into linear models. Substitution is not one of them. Substituting values leads to unsatisfactory results, particularly when more than one detection limit is present. Maximum likelihood estimates of slope and intercept produce a parametric regression model without resorting to substitution. MLE methods provide a best-fit line for data with one or more detection limits, assuming the residuals follow the chosen distribution. For a nonparametric approach, the Thiel-Sen line can be computed for censored data without an assumption of any specific distributional shape. One advantage of the Thiel-Sen line is that it can be computed when values for both x and y are censored. A third approach named logistic regression can be performed when the y variable is classified as either detect or nondetect, analogous to the contingency table process for testing group differences. This binary regression evaluates how the proportion of detects and nondetects changes as a function of one or more explanatory variables. Each of these methods is examined in turn.

### Substitution methods

By now there should be no doubt that substitution methods are inadequate. The example below demonstrates this in the context of regression.

Consider again the summer dissolved iron (DFe) data presented in Table 5 of Hughes and Millard (1988):

| DFe: | 20 | <10 | <10 | <10 | <10 | 7 | 3 | <3 | <3 |
|------|-----|------|------|------|------|------|------|------|------|
| Year: | 1977 | 1978 | 1979 | 1980 | 1981 | 1982 | 1983 | 1984 | 1985 |

To determine whether there is a trend in dissolved iron over time, a regression of DFe (y-variable) versus Year (x-variable) can be computed and the slope tested to determine if it is significantly different from zero. One analyst might set nondetects to the value of their detection limits, while another sets all nondetects to 0. The results for both are given below.

```
nondetects = dl
The regression equation is
DFe = 3508 - 1.77 YEAR

Predictor         Coef       SE Coef        T          P
Constant          3508.2       662.4       5.30      0.001
Year             -1.7667      0.3344      -5.28      0.001

S = 2.590      R-Sq = 80.0%        R-Sq(adj) = 77.1%


nondetects = 0
The regression equation is
DFeZero = 2215 - 1.12 YEAR

Predictor         Coef       SE Coef        T          P
Constant          2215         1627       1.36      0.215
Year             -1.1167      0.8211      -1.36      0.216

S = 6.360      R-Sq = 20.9%        R-Sq(adj) = 9.6%
```

When substituting the detection limit for all nondetects, the slope of DFe versus Year (−1.77) appears significantly different from zero with a t-test statistic of −5.28 and a p-value of 0.001. A significant trend of decreasing dissolved iron is declared. However when zeros are substituted for nondetects, the slope for Year (−1.12) is not significantly different from zero (p = 0.216), so no trend is found. The values for the intercept change by about one-third. The r-squared statistic changes from 80 to 21 percent. Neither equation can be considered better than the other. Neither is definitive. With only the evidence available from this procedure, neither equation is necessarily correct. Even if both had produced the same result of non-significance, the choice to substitute values somewhere in-between these two could produce a significant test result. Clearly substitution produces inadequate information on which to base any decision.

In contrast, survival analysis software can be used to compute regression equations for left-censored data. The result is a unique solution, with a defensible test for whether the slope coefficient differs from zero. Assuming a normal distribution for the residuals, MLE produces the estimates for slope and intercept below. These estimates are defensible as the best-fit parameters, given censored data and the assumptions of normality and linearity. The slope for Year (–1.73) has a p-value of essentially zero; a downtrend in summer iron concentrations occurs over this time period if a linear trend is assumed.

```
Coefficients estimated by mle
DFe = 3426 - 1.73 YEAR
Estimation Method:  Maximum Likelihood
Distribution:  Normal
```

Regression Table

| Predictor | Coef | Standard Error | Z | P | 95.0% Normal CI Lower | Upper |
|---|---|---|---|---|---|---|
| Intercept | 3426.1 | 859.3 | 3.99 | 0.000 | 1741.9 | 5110.2 |
| Year | -1.7260 | 0.4337 | -3.98 | 0.000 | -2.5760 | -0.8760 |
| Scale | 3.1083 | 0.9785 | | | 1.6771 | 5.7607 |

```
Log-Likelihood = -13.184
```

Thompson and Nelson (2003) compared the bias and precision of MLE regression to that of substituting one-half the detection limit and found that substitution produced biased estimates of slope, as well as producing confidence intervals that were too small. The errors they found were likely not as large as would be found in practice for environmental data, given the inconsistencies among laboratories in the determination of detection limits (see Chapter 3). Their study used simulations with consistently-defined limits. A more realistic setting would be to first censor at several higher detection limits for some proportion of the data, and then substitute fabricated numbers based on those limits. Errors from substitution in this more-realistic scenario would probably be considerably larger than those found in their study. Even with smaller errors, however, their study strongly advocated the method of maximum likelihood over substitution for performing regression with censored data.

## Maximum likelihood estimation

Maximum likelihood methods are used in a variety of disciplines to compute regression models. In reliability analysis, they are called "failure time models" (Meeker and Escobar, 1998). The response variable in that case is the time until a product fails, for example, until a light bulb burns out. In medical statistics, "accelerated failure time models" are used to predict the time until the recurrence of a disease, or until death (Collett, 2003). In economics, "censored maximum likelihood models" are used to predict the time until an event such as an interest rate change occurs, but have also been used with a response variable other than time. For example, Chay and Honore (1998) modeled incomes using MLE regression, where records were right-censored at tax-category ceilings. For the specific case of left-censored data that can include true zeros and whose residuals follow a normal distribution, MLE is sometimes called "Tobit analysis" after the economist James Tobin (Tobin, 1958). All of these methods are fundamentally identical. Terminology, as always, can be confusing.

For regression of right-censored "failure time models", uncensored observations are those where the subject's length of time, such as the time until death, is known exactly. The event has occurred and is recorded. Censored data are observations where the event has not yet happened by the time the experiment is finished. For these observations the time to occurrence is known only to be greater than some value. Regression analysis determines the significance of the effects of explanatory variables on the time until occurrence of the event – for example, does life expectancy increase with the amount of daily exercise?

In environmental science MLE can be used to model the response of left-censored variables other than time, such as concentrations. For example, Slymen et al. (1994) modeled the concentrations of the trace element tin in minnows as a function of exposure time to wastewaters. Twenty percent of tin concentrations were below detection limits. To be sure, there are environmental studies where the response variable is time, such as time to recovery of an ecosystem or time until death of a sentinel species organism. These can be modeled with MLE using standard right-censored failure-time analysis. However, the primary interest in this book is in using MLE for left-censored variables where something other than time is modeled.

MLE regression software must be able to incorporate left-censored observations using an "interval censored" or "arbitrary censored" format in order to correctly model variables that do not go negative, such as concentration. If this type of data entry is not allowed, the MLE procedure can be performed by flipping concentrations to a right-censored format prior to calculation, but there is one caution in doing so. The infinite upper bound for right-censored data will translate into an infinite lower bound, rather than a zero lower bound, for the original concentration variable. Parameters will be estimated assuming that negative concentrations are possible, and the resulting parameter estimates will be biased. However, if logarithms of concentration are used as the response variable, this bias will not be present. An infinite lower bound for logarithms will map into a lower bound of zero for the retransformed, original concentration units. Though flipping logarithms avoids biased regression parameters, using MLE software that allows direct input of left-censored data is less confusing, requiring fewer retransformations from flipped to unflipped scales, and so is highly recommended.

As a parametric method, hypothesis tests using MLE require that the data approximately follow an assumed distribution. For skewed environmental data where most variables have values spanning two or more orders of magnitude, a lognormal distribution is most often assumed. Parameter estimates for the slopes and intercept of a linear regression are computed by MLE in a manner similar to the description of MLE in Chapter 2. A likelihood function is written as

$$L = \prod p[e_i]^{\delta_i} \bullet F[e_i]^{1-\delta_i} \tag{12.1}$$

where

$\delta_i$ is the indicator of 0 for a censored observation and 1 for a detected observation,

$F[e_i]$ is the cumulative distribution function of the residuals, equaling Prob($e_i \le t$) for limit t,

$p[e_i]$ is the probability density function of the residuals, and

$e_i$ are the residuals from the regression equation

$$e_i = y_i - \sum \beta_j x_i \tag{12.2}$$

where the $\beta$s are the coefficients for the j explanatory variables.

The derivative of the log of L with respect to the $\beta$s is set to 0. Solving these equations produces the coefficients $\beta_j$ with the highest likelihood of matching the observed data, both censored and uncensored. The optimization is something like varying the regression surface (a line for one $\beta$, a plane for two $\beta$s, and a higher-dimensional surface in more than two dimensions) as it slices through the data until it finds the position with the smallest residual error.

Minitab® and many other statistics packages produce Wald's partial tests for the significance of each coefficient in MLE regression. Wald's tests estimate each coefficient along with an estimate of their standard error. The ratio of the coefficient to its standard error approximately follows a normal distribution, at least asymptotically (for large sample sizes) and when the distribution of the regression residuals follows the assumed distribution. Most textbooks on survival analysis recommend using likelihood ratio tests (see Chapter 2) instead of Wald's tests, due to the uncertainty in how quickly the Wald's ratios converge on their true value for small data sets that only approximately fit the assumed distribution. Likelihood ratio tests still require that data follow the assumed distribution, but unlike Wald's tests, their validity does not depend on whether the coefficients themselves follow a normal distribution.

The test for determining the overall significance of an MLE regression model, similar in concept to the overall F-test in least-squares regression, is the overall likelihood-ratio test. This test determines whether the entire model being tested is an improvement over using no model at all, that is, over the "null model" where all $\beta$s equal 0. The overall test statistic is:

$$G^2{}_0 = 2 \, [\ln L(\beta) - \ln L(0)] \tag{12.3}$$

$$= [-2 \ln L(0)] - [-2 \ln L(\beta)]$$

where lnL(0) represents the log-likelihood of the null model. Statistical software that does not report $G^2_0$ will report values either for the log-likelihood, lnL($\beta$), or the "$-2$ log-likelihood" $-2$lnL($\beta$). Minitab® prints the log-likelihood lnL($\beta$) for each regression model. When it is not reported, $G^2_0$ can be computed by obtaining the log-likelihood for the null model. In Minitab® this is done by using the

### Reliability/Survival > Parametric distribution analysis

command, without any explanatory variables, as was done for estimating the mean in Chapter 6. The test statistic $G^2_0$ is compared to a chi-square distribution with k degrees of freedom, where k is the number of explanatory variables in the model, to determine a p-value for the overall test.

Partial likelihood tests for the $\beta$ coefficients of each explanatory variable in the model are alternatives to the Wald's tests computed by most statistical software. They are similar in concept to the partial t-tests of least-squares regression. If log-likelihoods for each explanatory variable are not printed by the software, they can be obtained in a manner similar to equation 12.3 by running the MLE twice, once with and once without the explanatory variable to be tested. The test statistic is twice the difference in log-likelihoods between the two models (equation 12.4), measuring how the fit to the data is improved by the use of that variable:

$$G^2_{partial} = 2 \, [\text{lnL}(\beta_{with}) - \text{lnL}(\beta_{without})] \qquad (12.4)$$

$$= [-2 \, \text{lnL}(\beta_{without})] - [-2 \, \text{lnL}(\beta_{with})]$$

A p-value is obtained by comparing the test statistic to a chi-square distribution with one degree of freedom. If the p-value is less than the significance level $\alpha$, the null hypothesis that the coefficient $\beta$ equals 0 is rejected, and including the explanatory variable provides a significant improvement in model fit. If the p-value is large and the null hypothesis is not rejected, then the variable can be dropped from the list of useful predictor variables.

### *An example of MLE regression*

Are TCE concentrations in ground water a function of population density, landuse, or depth to the water surface? The data set TCEReg contains information on TCE concentrations in ground waters of Long Island, NY, along with data on several explanatory variables (Eckhardt et al., 1989). Of interest is determining whether any of the explanatory variables significantly affect TCE concentrations, and if so, estimating their slopes.

The relationship between TCE concentration and the three explanatory variables – %IndLU (percent industrial land use), Depth, and Popden – is measured using MLE regression. The column TCEConc contains the concentrations and detection limits for TCE. There are multiple detection limits. The column BDL0 is the indicator column for TCE, having a value of 0 for data below detection limits (hence the name) and a 1 for detected observations. MLE software for left-censored or arbitrarily-censored data requires a start and an end column, the interval endpoints format of Chapter 3. The TCEConc column is the end column, containing the upper limit value of the detection limit for censored observations. A start column must be created by multiplying TCEConc times BDL0. The result is stored in the column TCE0. In TCE0 all nondetects are represented by a value of 0. All detected observations have the same TCE concentration in both the start and end columns. A few example entries for these data are given below. As an example, the last entry is a detected concentration of 1 µg/L, the smallest possible detected concentration for the time period when the detection limit was 1.

| BDL0 | TCEConc | LU | Popden | %IndLU | Depth | TCE0 |
|------|---------|-----|--------|--------|-------|------|
| 0 | 1.000 | 9 | 9 | 10 | 103 | 0.000 |
| 0 | 1.000 | 9 | 11 | 0 | 32 | 0.000 |
| 0 | 1.000 | 8 | 3 | 4 | 142 | 0.000 |
| 1 | 32.00 | 9 | 11 | 3 | 69 | 32.000 |
| 0 | 1.000 | 9 | 11 | 0 | 32 | 0.000 |
| 1 | 13.00 | 9 | 14 | 7 | 89 | 13.000 |
| 0 | 1.000 | 5 | 6 | 1 | 232 | 0.000 |
| 1 | 150.0 | 8 | 6 | 6 | 177 | 150.000 |
| 1 | 1.000 | 8 | 3 | 4 | 207 | 1.000 |

The maximum likelihood equation is solved by using the

Stat > Reliability/Survival > Regression with Life Data

command of Minitab®. With start variable TCE0 and end variable TCEConc, and the three explanatory variables entered into the Model window, the output from MLE regression assuming a normal distribution is:

```
Estimation Method: Maximum Likelihood
Distribution:   Normal


Regression Table
```

| Predictor | Coef | Error | Z | P | Lower | Upper |
|-----------|------|-------|---|---|-------|-------|
| Intercept | 6.31118 | 4.55867 | 1.38 | 0.166 | -2.62366 | 15.2459 |
| Popden | 0.373907 | 0.538686 | 0.69 | 0.488 | -0.681899 | 1.42970 |
| %IND LU | 0.0791126 | 0.439033 | 0.18 | 0.857 | -0.781377 | 0.939602 |
| Depth | -0.0091394 | 0.0129492 | -0.71 | 0.480 | -0.0345193 | 0.0162406 |
| Scale | 30.3921 | 1.36742 | | | 27.8267 | 33.1939 |

```
Log-Likelihood = -1069.059
```

The probability plot of residuals for the regression (Figure 12.1) indicates that a transformation is necessary. The residuals do not match the straight line representing the normal distribution.

A lognormal distribution is assumed for the residuals and the procedure run again. The residuals are essentially linear (Figure 12.2), so that a lognormal distri-

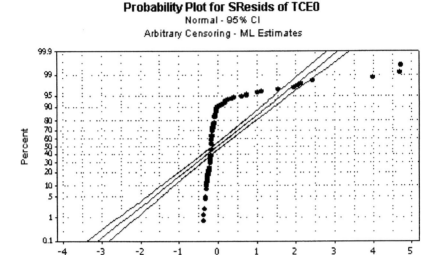

**Figure 12.1.** Probability plot of residuals for MLE regression of the TCE data, assuming a normal distribution.

bution is suitable for MLE regression of the TCE data. The output from the procedure is given below.

```
Estimation Method: Maximum Likelihood
Distribution:    Lognormal


Regression Table
```

| Predictor | Coef | Standard Error | Z | P | 95.0% Normal CI Lower | Upper |
|-----------|------|----------------|---|---|------|-------|
| Intercept | -2.88026 | 0.823547 | -3.50 | 0.000 | -4.49438 | -1.26613 |
| Popden | 0.250904 | 0.0745204 | 3.37 | 0.001 | 0.104846 | 0.396961 |
| %IND LU | 0.0406455 | 0.0526390 | 0.77 | 0.440 | -0.0625251 | 0.143816 |
| Depth | -0.0043726 | 0.0023329 | -1.87 | 0.061 | -0.0089450 | 0.0001998 |
| Scale | 2.81166 | 0.311129 | | | 2.26345 | 3.49264 |

```
Log-Likelihood = -302.931
```

To compute the overall test for whether this three-variable model predicts TCE concentrations better than simply the mean concentration (the null model), the log-likelihood for no explanatory variables is calculated using the

Stat > Reliability/Survival > Distribution Analysis (Arbitrary censoring) > Parametric Distribution Analysis

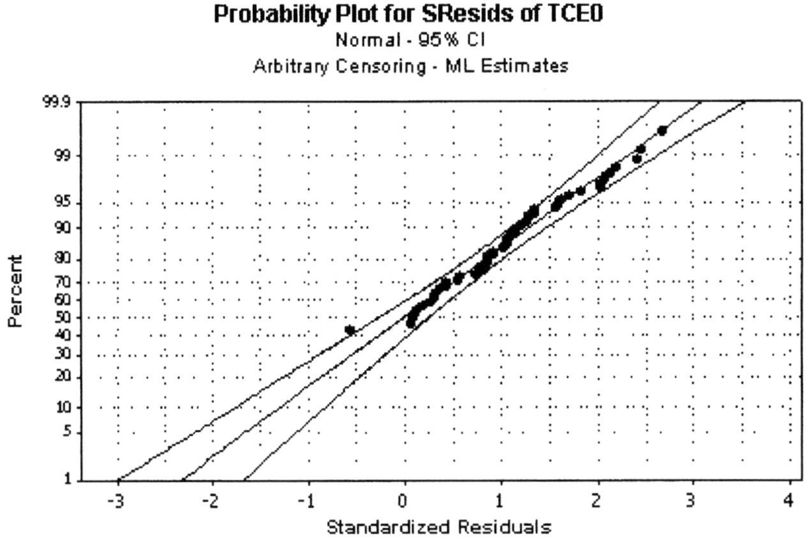

**Probability Plot for SResids of TCE0**
Normal - 95% CI
Arbitrary Censoring - ML Estimates

**Figure 12.2.** Probability plot of residuals for MLE regression of the TCE data, assuming a lognormal distribution.

command. Assuming a lognormal distribution, the null model produces a log-likelihood of −316.404:

```
Estimation Method: Maximum Likelihood
Distribution:      Lognormal

Parameter Estimates
```

| Parameter | Estimate | Standard Error | 95.0% Normal CI Lower | Upper |
|-----------|----------|----------------|-------|-------|
| Location | -1.77893 | 0.415959 | -2.59420 | -0.963676 |
| Scale | 2.93033 | 0.327988 | 2.35311 | 3.64913 |

```
Log-Likelihood = -316.404
```

Using equation 12.3, the test statistic $G^2{}_0$ is computed for the overall test as:

$$G^2{}_0 = 2 \ [\ln L(\beta) - \ln L(0)] = 2 \ [-302.931 - (-316.404)] = 26.95$$

Comparing 26.95 to a table of the chi-square distribution with k=3 degrees of freedom (3 explanatory variables), the resulting p-value equals < 0.001, less than the alpha of 0.05. So the three-variable model is considered better than no model at all.

The Wald's tests (Z statistics) for coefficients of the three explanatory variables indicate previously that Popden is a significant predictor of TCE concentration (p = 0.001), that %IndLU is not (p = 0.440), and that Depth is on the edge with a p-value of 0.06. Since the Wald's tests are believed by many to be only approximate, a likelihood-ratio test can be conducted for a variable such as Depth whose Wald's test conclusion might be in doubt. To do this, the lognormal MLE regression is again computed, this time without Depth as an explanatory variable. The output for this two-variable regression model is below.

```
Distribution:  Lognormal
Relationship with accelerating variable(s):  Linear, Linear
```

Regression Table

| Predictor | Coef | Standard Error | Z | P | 95.0% Normal CI Lower | Upper |
|---|---|---|---|---|---|---|
| Intercept | -3.82240 | 0.767808 | -4.98 | 0.000 | -5.32727 | -2.31752 |
| Popden | 0.300775 | 0.0740682 | 4.06 | 0.000 | 0.155604 | 0.445946 |
| %IndLU | 0.0376751 | 0.0529982 | 0.71 | 0.477 | -0.0661996 | 0.141550 |
| Scale | 2.83152 | 0.314002 | | | 2.27838 | 3.51896 |

```
Log-Likelihood = -305.035
```

Following equation 12.4 the partial log-likelihood test for Depth is computed as:

$$G^2{}_{partial} = 2\,[\ln L(\beta_{with}) - \ln L(\beta_{without})] = 2[-302.931 - (-305.035)] = 4.208$$

with an associated p-value from a chi-square distribution with 1 degree of freedom of 0.04. This p-value is smaller than for the Wald's test (and the log-likelihood test is to be preferred), so that Depth is considered a significant variable and should be retained in the model. Based on the partial log-likelihood tests, the best regression model for these data has Popden and Depth as explanatory variables. The final model for explaining TCE concentrations is

lnTCE = −2.79 + 0.260*Popden − 0.004*Depth

```
Distribution:  Lognormal
Relationship with accelerating variable(s):  Linear
```

Regression Table

| Predictor | Coef | Standard Error | Z | P | 95.0% Normal CI Lower | Upper |
|---|---|---|---|---|---|---|
| Intercept | -2.79066 | 0.810181 | -3.44 | 0.001 | -4.37859 | -1.20273 |
| POPDEN | 0.259589 | 0.0740544 | 3.51 | 0.000 | 0.114445 | 0.404733 |
| DEPTH | -0.0043407 | 0.0023406 | -1.85 | 0.064 | -0.0089282 | 0.0002468 |
| Scale | 2.81474 | 0.311546 | | | 2.26582 | 3.49665 |

```
Log-Likelihood = -303.227
```

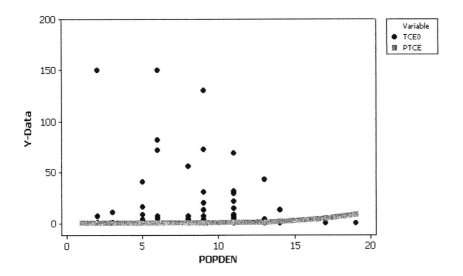

**Figure 12.3.** TCE concentrations as a function of population density (Popden). The lognormal MLE regression model is shown as the curve.

The MLE lognormal regression model is pictured as a curve in Figure 12.3. Most of the data are nondetects, and plot on top of one another at the low end of the y-axis scale. A variable which would improve on this model is one that would explain the high concentrations occasionally seen at lower population densities.

### Theil-Sen nonparametric regression

A nonparametric alternative to the estimate of slope in linear regression is the median of all possible slopes between pairs of data. Suggested first by Theil (1950), Sen (1968) placed the estimator in the context of being a Hodges-Lehman estimator, determining its confidence interval and establishing it as a robust alternative to the least-squares slope of ordinary linear regression. Hirsch and Slack (1984) used the Theil-Sen method to estimate the trend slope for the Seasonal-Kendall test, which has become one of the most popular tests in environmental studies for determining changes over time. The Theil-Sen slope and confidence interval are related to Kendall's tau correlation coefficient (see Chapter 11). The test for whether the Theil-Sen slope is significantly different from zero is also the test for whether Kendall's tau is significantly different from zero. Neither relies on an assumption of normality for their validity. If the trend $\Delta y / \Delta x$ measured by the Theil-Sen slope is subtracted from the y variable, the correlation between the residuals and the x variable will have a tau coefficient of zero.

For censored data, a Theil-Sen estimate of slope can be calculated in one of two ways, each related to the properties of the Theil-Sen slope for uncensored data. The first slope estimate is the median of all possible slopes between pairs of data, given that some of those slopes will be interval-censored (within a range of values). The second slope estimate is one that when subtracted from the y variable most closely produces a tau correlation coefficient of zero between the residual and the x variable. These two slope estimates need not be identical. The first method is implemented in the Minitab® macro ckend.mac, and the second is implemented in the ats.mac macro. The two estimators are discussed below. One attractive property of either Theil-Sen estimator is that it can be computed for doubly-censored data, data where x and y are both censored. The more traditional estimators of slope for censored data, including Buckley-James regression and MLE methods, allow only the y variable to be censored.

*Median of pairwise slopes*

Censored concentration data have values within an interval between 0 and the reporting limit. The endpoints of this interval can be used when computing possible slopes to determine a range that covers the possible slopes for each censored data point. Consider a simple example with three data points:

| Point | X | Y |
|-------|---|-----|
| 1 | 2 | <1 |
| 2 | 5 | 4 |
| 3 | 8 | 6 |

There are three slopes that can be computed between these three pairs of points.

Between points 2 and 1: $\dfrac{4-<1}{5-2} = \dfrac{4-1}{5-2} to \dfrac{4-0}{5-2} = \dfrac{3}{3} to \dfrac{4}{3}$

Between points 3 and 1: $\dfrac{6-<1}{8-2} = \dfrac{6-1}{8-2} to \dfrac{6-0}{8-2} = \dfrac{5}{6} to \dfrac{6}{6}$

Between points 3 and 2: $\dfrac{6-4}{8-5} = \dfrac{2}{3}$

Only the slope between the two uncensored points is known uniquely. The other two slopes lie within a defined range. The median of these three slopes is the first method of calculating the Theil-Sen slope estimate. But how is a median to be computed from interval-censored data?

Kaplan-Meier estimates of the median or other percentiles can be computed for right-censored data. But there is more information in the intervals than stating that the first slope, for example, is a right-censored value greater than 3/3. There is also an upper end to the estimated range in possible slopes. Both endpoints need to be recognized. This can be done with a survival analysis method that computes nonparametric estimates of percentiles when data are interval-censored (Turnbull, 1976). Turnbull's method iteratively estimates the survival function in a process similar to Kaplan-Meier when the censored variable (the "survival time") includes both the start and end points of an interval. For details on the computation, see Klein and Moeschberger (2003). Turnbull's method is available in most survival analysis software, and is implemented in the ckend macro to provide the median slope of paired (x,y) data that include nondetects. The test of significance for whether this slope is significantly different from zero is the significance test for Kendall's tau, also computed in the macro.

As an example, a slope for predicting the TCE concentrations previously modeled using MLE regression can be computed with this nonparametric method. The ckend macro is run using TCE as the censored y variable and population density as the x variable. A zero in the BDL0 indicator column designates a censored entry for TCE. The macro is invoked with four arguments, X, Y, and the censoring indicators for X and Y:

```
%ckend c4 c2 c7 c1;
SUBC> cens 0.

Executing from file: ckend.MAC

Kendall's tau
S         3703.00
tau       0.121885
taub      0.244847
z         3.26106
pval      0.00110993

The median slope is between 0.333333 and 0.800000
Turnbull estimate of median slope
slope     0.555556
```

To review what is produced by the macro, S is the number of concordant data pairs minus the number of discordant pairs, and is zero when there is no monotonic association between Y and X when the null hypothesis is true. As S increases in absolute value, the evidence for correlation increases. Tau is S scaled to a value between −1 and +1. Taub is Kendall's tau-b, described in Chapter 11, which measures the correlation only for pairs of data whose X or Y values are not tied. The test statistic z, compared to a standard normal distribution with the resulting p-value "pval", tests whether the correlation coefficient, and therefore the slope, is significantly different from zero.

The Turnbull estimate of median slope for these data is 0.56, significantly different from zero with a p-value of about 0.001. This indicates that TCE concentrations increase at a rate of 0.56 µg/L for every unit increase in population density. Though the nonparametric test makes no assumption about the distribution of the residuals of the data, computing a slope implies that the data follow a linear pattern. If the data do not follow a linear pattern, either the y or x variables should be transformed to produce one before a slope is computed. Otherwise the statement that a single slope describes the change in value for data in the original units is not correct. Logarithms of TCE were previously used to improve the linear relationship with population density. After taking the natural logs of the TCE concentrations, the ckend macro for lnTCE versus population density produces:

```
Kendall's tau
S          3703.00
tau        0.121885
taub       0.244847
z          3.26106
pval       0.00110993


The median slope is between 0.121808 and 0.231049
Turnbull estimate of median slope
slope      0.173287
```

Note that the tau correlation coefficient is the same for both original and log data, but the slope has changed. The equation indicates that TCE concentrations increase by 0.173 natural log units for every unit increase in population density. This translates into an average increase of 18.5% per year, using the formula

$$\text{percent change in Y per year} = (e^{b_1} - 1) \bullet 100 \qquad (12.5)$$

where $b_1$ is the slope in natural log units.

*The Theil-Sen slope producing a zero value for tau (Akritas–Theil-Sen)*

Akritas et al. (1995) extended the Theil-Sen slope to censored data by calculating the slope that, when subtracted from the y data, would produce an approximately zero value for Kendall's tau correlation coefficient. They found that this method had lower bias and standard error than several alternatives, including a weighted least-squares approach and a median of pairwise slopes method (though it was not the Turnbull method of the ckend macro). A later study by Wilcox (1998) showed that the Akritas-Theil-Sen slope had a "substantial advantage" in bias and precision over Buckley-James regression, the most commonly used nonparametric regression method for censored data (see the later section on "Additional methods for censored regression"). The Akritas-Theil-Sen method has as much utility for trend analysis and other regression models for censored data as does the original Theil-Sen slope for uncensored data.

To compute the Akritas-Theil-Sen slope estimator, set an initial estimate for the slope, subtract this from the y variable to produce the y-residuals, and then determine Kendall's S statistic between the residuals and the x variable. Next, conduct an iterative search to find the slope that will produce an S of zero. Because the distribution function of the test statistic S is a step function, there may be more than one slope that will produce a value of zero S. Therefore the final Akritas-Theil-Sen slope is considered to be the one halfway between the maximum and minimum slopes that produce a value of zero for S.

Though both approaches to computing a Theil-Sen slope provide estimates for doubly-censored data, they do so using different methods. Only slopes with uncensored x values are used in computing the Akritas-Theil-Sen estimator. Other slope estimates (interval estimates of slope due to a censored x value) are ignored. Kendall's S is also computed only for data with uncensored x values by the Akritas-Theil-Sen method. In contrast, the Turnbull estimate of median slope and the corresponding (standard) calculation of Kendall's S considers all possible slopes, even those where the x values are censored. If all x data are uncensored the two estimates of S and tau will be identical. Otherwise, they are similar but not exactly the same.

As an example, a nonparametric slope estimate of TCE concentration as a function of population density (Popden) is computed using the Akritas-Theil-Sen macro ats.mac. The same four arguments used for the ckend macro are required: X, Y, and censoring indicators for X and Y.

```
%ats    c4 c2 c7 c1;    cens 0.
```

The subcommand cens is used because the indicator for censored data is a 0 in this data set, rather than the default of 1. The resulting output includes the slope which, when multiplied by Popden and subtracted from the TCE concentration data, results in residuals having a Kendall's tau correlation of zero:

```
A-T-S line
stau      3703
tau       0.122

Slope         0.384
Intercept    -4.218
```

Software for these two methods is not readily available in standard statistics packages. The Turnbull estimator of percentiles for interval-censored data is available in most survival analysis packages, so that implementing the median of pairwise slopes procedure is relatively easy once all the possible pairwise slopes, including those that are interval-censored, have been calculated. The ATS procedure can programmed using an iterative search process.

*Nonparametric estimates of intercept*

Several possible nonparametric estimates for an intercept to accompany the Theil-Sen slope have been evaluated for uncensored data (Dietz, 1987; Dietz, 1989; Hollander and Wolfe, 1999, section 9.4). However, their use has not been explicitly evaluated for the case of censored data. Dietz (1987; 1989) found that the median residual (equation 12.6) was a relatively efficient measure of the intercept for a linear equation based on the Theil-Sen slope $b_{TS}$.

$$\hat{b}_0 = median[Y_i - b_{TS} \bullet X_i] \text{ for } i = 1,\dots n \tag{12.6}$$

A second estimator had slightly lower mean squared error under specific circumstances. It was the median of all pairwise (Walsh) averages of residuals, in the Hodges-Lehmann class of estimators (equation 12.7):

$$\hat{b}_{HL} = median\left(\frac{[Y_i - b_{TS} \bullet X_i] + [Y_j - b_{TS} \bullet X_j]}{2}\right) \text{ for } i,j = 1,...n \text{ and } j \neq i \quad (12.7)$$

A third estimator was attributed to W.J. Conover (see Conover, 1999). It had higher mean square error but is simpler to compute. Remember from basic statistics that the least-squares regression line goes through the point ($\overline{X}, \overline{Y}$). The Theil-Sen line can be placed through the median of X ($X_{med}$) and median of Y ($Y_{med}$) by using the intercept in equation 12.8:

$$\hat{a} = Y_{med} - b_{TS} \bullet X_{med} \quad (12.8)$$

where $b_{TS}$ is the Theil-Sen slope estimator. This is the form of the trend line in the Seasonal Kendall trend analysis process of Hirsch and Slack (1984). For censored data, it would require that the median of both X and Y variables be computed by Kaplan-Meier (see Chapter 6) or another method appropriate for censored data. Additional information on these three estimates of intercept using uncensored data is found in Hettsmansperger et al. (1997).

The median residual intercept of equation 12.6 was chosen for both the ckend and ats macros, as it is more efficient than equation 12.8 and simpler to compute for censored data than is equation 12.7. Note that though Akritas et al. (1995) derived the slope estimate used in the ats macro, their paper looked only at slope estimates and did not evaluate any corresponding estimates for intercept. Until a study of intercept terms is conducted specifically for censored data, the uncensored results favoring equations 12.6 or 12.7 is all that is available. Censored Y observations produce interval-censored residuals, and so to solve the equivalent of equation 12.6 the median residual is computed using a Turnbull estimate.

The result of using any of these intercept estimates, along with the Theil-Sen slope estimate, is a line less strongly affected by outliers than is regression that assumes a normal error distribution, such as MLE regression. A line using the Theil-Sen slope predicts the conditional median of Y (the median of Y for any given X), rather than the conditional mean of Y provided by either least-squares or MLE methods. If the residuals follow a normal distribution, the Theil-Sen line and the normal-theory line should be quite similar. For linear patterns with non-normal residuals or influential outliers, the Theil-Sen line should more closely slice through the center of observations. A comparison of MLE and Theil-Sen results for fitting a straight line to the logarithms of the TCE data is listed in Table 12.1.

**Table 12.1.** Comparison of slopes and intercepts for three straight lines fit to the natural logarithms of TCE concentrations. The single explanatory variable is population density.

| Method | Slope | Intercept | p-value |
|---|---|---|---|
| MLE (lognormal) | 0.309 | −3.73 | 0.000 |
| Theil-Sen median of possible slopes | 0.173 | −1.903 | 0.001 |
| ATS (slope for tau = 0) | 0.383 | −4.22 | 0.001 |

The three lines are plotted in Figure 12.4. Slopes and intercept for the MLE and ATS lines are quite similar. This is expected when residuals are approximately normal, as they are after taking logarithms of TCE concentrations. Remember that most of the data are nondetects, with the proportion of nondetects decreasing as population density increases. The three lines are all sensitive to those data.

Theil-Sen estimates for multivariate regression have been discussed in a few papers but not yet implemented in software for routine use. Dietz and Killeen (1981) described a multivariate application of the Theil-Sen slope to test for trend in uncensored data. Akritas and Siebert (1996) derived Kendall's tau for partial correlation with censored data, leading to possibilities of implementing a Kendall and Theil-Sen approach for multivariate censored data. However that possibility has not yet been realized in available software.

## Logistic regression

The regression methods previously discussed used concentration as their response variable. With logistic regression, the y variable is a probability $\pi$, such as the probability of the concentration being above the detection limit. The probability of falling below the detection limit is then $1-\pi$ (the Greek letter $\pi$ is used here to avoid using the letter p and so avoid confusing the probability of detection with a p-value). Logistic regression models the probability $\pi$ as a function of the effects of one or more explanatory variables. Observations for the response variable are recorded as belonging either in one category or another, such as below or above a detection limit. The regression equation predicts the probability of falling into one category. In environmental studies, logistic regression can be used when the observed data are so frequently below the detection limit that it is difficult to determine whether the data follow any particular distribution, or whether the change in concentration is linear for any particular units of y or x.

Logistic regression can be written in a form that appears much like least-squares regression:

$$\ln\left(\frac{\pi}{1-\pi}\right) = b_0 + b_j X_j \tag{12.9}$$

Logistic regression relates the logistic (or logit) transformation of the y-variable (the left side of equation 12.9) to a linear function of the x variable(s). The right side of equation 12.9 looks similar to a least-squares multiple regression equation, where $X_j$ represents a vector of j = one or more explanatory variables, and the $b_j$ are the fitted slope coefficients. The left side of the equation is called the logit or logistic transform, the natural logarithm of the odds ratio $\frac{\pi}{1-\pi}$ for the occurrence of an event. If $\pi = 0.8$ or four-fifths, then the odds ratio for the event is $\frac{0.8}{0.2}$, or 4 to 1.

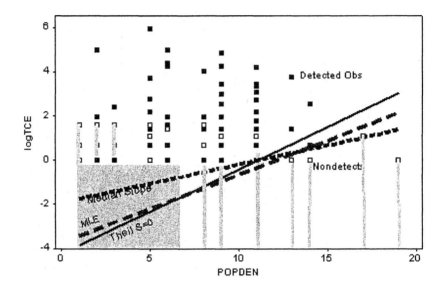

**Figure 12.4.** Three linear models for log of TCE concentrations versus population density. Data are from Eckhardt (1989).

The logistic transform in equation 12.9 can be solved for the probability $\pi$, producing equation 12.10.

$$\pi = \frac{\exp(b_0 + b_j X_j)}{1 + \exp(b_0 + b_j X_j)} \qquad (12.10)$$

For a unit increase in the $j^{\text{th}}$ explanatory variable $X_j$, the odds ratio $\frac{\pi}{1-\pi}$ increases by a multiplicative factor of $e^{b_j}$. When $\pi$ is plotted versus an explanatory variable as in Figure 12.5, the result is an S-shaped curve. The S-curve is flexible, becoming almost linear at the central portion near $\pi = 0.5$ while changing value much more slowly near the extremes of $\pi = 0$ or 1. At the inflection point of $\pi = 0.5$ (probability of detection equals 50%), the logit function at the left side of equation 12.9 equals $\ln\left(\frac{0.5}{1-0.5}\right)$, or 0.

The S-curve that best fits observed frequencies of detection is determined by maximum likelihood. Equation 12.11 is the log-likelihood equation for i = 1 to n observations where $\pi$ is the probability of a detect.

$$L = \sum_{i=1}^{n} \left( y_i \bullet \ln[\pi] + (1 - y_i) \bullet \ln[1 - \pi] \right) \qquad (12.11)$$

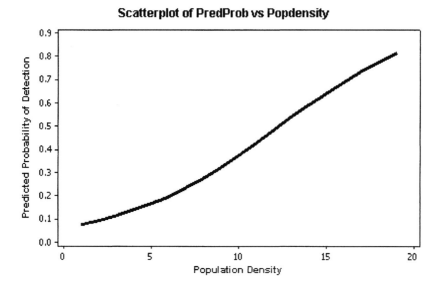

**Figure 12.5.** Logistic regression model of TCE detection. Note the S shape of the function.

The y value for each observation is either a 0 (nondetect) or a 1 (detect). For a nondetect, the left side of the expression inside the summation sign is zero and the right side becomes $\ln(1-\pi)$. This will be a maximum of $\ln(1)$ for $\pi$ at 0, which fits the observed value the best. For a detected observation $y = 1$, the right side becomes zero and the left side becomes $\ln(\pi)$. This will fit the observed data best at $\ln(1)$ when $\pi = 1$. Maximizing the log-likelihood function is a search for values of the intercept $b_0$ and the slopes $b_j$ that produce a value for L closest to $\ln(1) = 0$. The solution maximizes the fit between the estimated probabilities, which are a function of the j explanatory variables $X_j$, and the observed data.

A single log-likelihood value tells the investigator very little. Instead, what is informative are comparisons between two or more nested models. The two models are compared using a likelihood-ratio test, to determine which model is preferred – which explanatory variables should be included in the logistic regression equation. The test for determining the overall significance of a logistic regression model, similar in concept to the overall F-test in least-squares regression, is the overall likelihood-ratio test. This test determines whether the entire model is an improvement over using no model at all, that is, over the "null model" where the slope coefficients for all explanatory variables equal 0. For the null model, the best estimate of $\pi$ is the average proportion of detections, regardless of the value of X. The overall test statistic is:

$$G^2{}_0 = 2 \left[\ln L(\beta) - \ln L(0)\right] \tag{12.12}$$

$$= \left[-2 \ln L(0)\right] - \left[-2 \ln L(\beta)\right]$$

where $\beta$ represents the population slope coefficients being estimated, $\ln L(\beta)$ is the log-likelihood of the tested model and $\ln L(0)$ is the log-likelihood of the null model. Statistical software will report values either for the log-likelihood or the "$-2$ log-likelihood" $-2\ln L(\beta)$. Minitab® prints the log likelihood for each regression model, along with the overall test statistic $G^2{}_0$. The test statistic $G^2{}_0$ is compared to a chi-square distribution with k degrees of freedom, where k is the number of explanatory variables in the model, to determine a p-value for the overall test.

Note that while Minitab® and most other software computes $\pi$ as the probability of observing a value of 1, SAS® considers $\pi$ as the probability of observing a 0. The results in one direction can be directly converted to the other, yet confusion results from not knowing which direction is being used. Hypothesis tests results (log-likelihood statistics and p-values) will be identical, but the slope coefficients themselves will have the opposite sign. Estimates for $\pi$ will equal $1-\pi$ for the other direction. Estimated odds ratios will be the inverse of the other (3 to 1 versus 1 to 3). If your software is computing logistic regression in the opposite direction from what you expect, the easiest remedy is to reverse the binary 0/1 assignment of the response variable and run the procedure again.

*The TCE Example, continued*

Is the probability of detecting TCE in the shallow groundwaters of Long Island, NY related to population density or depth to the water table? The 247 observations from Eckhardt et al. (1989) are found in TCEReg.xls. Logistic regression is run in Minitab® using the command

### Stat > Regression > Binary Logistic Regression

The original data have several detection limits, so a binary response variable LT5_0 was calculated to record whether or not each observation was above (1) or below (0) the maximum detection limit of 5 µg/L. The overall likelihood-ratio test determines whether the two variables Popden and Depth together significantly effect the probabilities of TCE concentrations exceeding 5. If not, the best estimate of $\pi$ for all depths and population densities is the average exceedance probability of 29/247 or 11.7%. The results are printed below.

```
Binary Logistic Regression: LT5_0 versus Popden, Depth
Link Function: Logit
Response Information
LT5_0          1              29        (Event)
               0             218
          Total             247
```

```
Logistic Regression Table
                Odds                                        95% CI
Predictor    Coef        SE Coef      Z        P      Ratio  Lower  Upper
Constant    -2.92781    0.546633    -5.36    0.000
Popden       0.161926   0.0523220    3.09    0.002    1.18   1.06   1.30
Depth       -0.0011198  0.0016968   -0.66    0.509    1.00   1.00   1.00

Log-Likelihood = -82.954
Test that all slopes are zero: G = 12.787, DF = 2, P-Value =
0.002
```

The (Event) designation for the value of 1 on the output shows that Minitab® is computing the probabilities of observing a 1. The log-likelihood $L(\beta)$ of this model equals $-82.95$. $G^2{}_0$ is reported as 12.79, which is significant at a very low p-value of 0.002. Therefore the two variables Popden and Depth together significantly affect the probabilities of TCE exceeding 5 µg/L in groundwater.

The other tests for significance that are important in logistic regression are partial tests, which determine whether the effect of each independent variable on $\pi$ is significant. Partial tests are used for model building, determining which explanatory variables to keep in the model and which to ignore. Two methods for computing partial tests have been used in logistic regression – likelihood-ratio tests and Wald's tests. Wald's tests compute an estimate of the slope for each explanatory variable, and an estimate of the standard error of that slope. Test statistics are commonly printed out in computer software in a similar location to partial t-tests in linear regression (though Wald's statistics do not follow a t-distribution). The Wald's ratio of slope to its standard error is approximately normally distributed. The square of the ratio is approximately distributed as a chi-square statistic. Either the normal or chi-square form of the test is used to produce a p-value. Using the normal distribution form of the test, Minitab® reports a partial Wald's test for Popden that is significant at a p-value of 0.002. The partial test for Depth is not significant (p = 0.509). Therefore a model with Popden as the sole explanatory variable would be preferable to the two-variable model above. Hosmer and Lemeshow (2000) state that Wald's estimates of standard error are unreliable for all but large sample sizes, with test statistics that are often too small, resulting in p-values larger than they actually should be. They recommend using partial likelihood-ratio tests instead of Wald's tests.

A partial likelihood-ratio test for each $\beta$ coefficient can be obtained by running the MLE twice, once with and once without the explanatory variable. The difference in log-likelihoods measures how the fit to the data is improved by the use of that variable. A p-value for the test is computed by comparing twice the difference in log-likelihoods to a chi-square distribution with one degree of freedom (equation 12.13):

$$G^2{}_{partial} = 2\ [\ln L(\beta_{with}) - \ln L(\beta_{without})] \qquad (12.13)$$
$$= [-2\ \ln L(\beta_{without})] - [-2\ \ln L(\beta_{with})]$$

If the p-value is less than the significance level $\alpha$, the null hypothesis that the coefficient $\beta$ equals 0 is rejected, and the explanatory variable provides a significant improvement in model fit. If the p-value is large and the null hypothesis is not rejected, the variable is dropped from the list of useful predictor variables.

To use a log-likelihood test for whether Depth significantly affects the exceedance probability of TCE, the logistic regression model is run using Popden as the only explanatory variable, and its log-likelihood statistic recorded.

```
Binary Logistic Regression: LT5_0 versus Popden
Link Function: Logit
Response Information
Variable   Value   Count
LT5_0        1        29        (Event)
             0       218
          Total      247
```

```
Logistic Regression Table
                                          Odds      95% CI
Predictor    Coef      SE Coef     Z      P    Ratio  Lower  Upper
Constant   -3.16995   0.428778  -7.39  0.000
Popden      0.173924  0.0500166  3.48  0.001  1.19   1.08   1.31
```

```
Log-Likelihood = -83.193
Test that all slopes are zero: G = 12.309, DF = 1,
P-Value = 0.000
```

The log-likelihood without Depth as an explanatory variable is $-83.193$. Comparing this to the log-likelihood for the model with both Popden and Density, the partial likelihood-ratio test statistic for Depth is:

$$G^2_{partial} = 2 (-82.954 - [-83.193]) = 0.478$$

From Minitab's®

### Calc > Probability Distributions > chi-square

command, the probability of exceeding 0.478 in a chi-square distribution with 1 degree of freedom is 0.49. Because this p-value is much larger than an alpha of 0.05, the null hypothesis of no effect is not rejected; Depth is adding little to the model and can be dropped. Similarly, for a log-likelihood test of the effect of Popden, that variable is dropped from the two-variable model. With Depth as the only explanatory variable, a log-likelihood of $-87.954$ results. Subtracting from the log-likelihood of $-82.954$ when both variables were in the model, and multiplying by 2, produces a test statistic $G^2_{partial}$ of 10.0 and a p-value of 0.0016. The effect of Popden is significant, and should remain in the model. Note that for this relatively large data set of 247 observations the results of the Wald tests and log-likelihood tests, as measured by their p-values, are essentially the same.

*Likelihood r-squared*

An overall measure of the 'strength' of the regression relationship for logistic regression is the likelihood r-squared (equation 11.3, repeated here as equation 12.14):

$$\text{likelihood } r^2 = 1 - \exp\left(-\frac{G_0^2}{n}\right) \quad\quad (12.14)$$

where $G_0^2$ is the overall likelihood ratio test statistic for the comparison between the selected model and the null model. For the TCE exceedance data with Popden as the sole explanatory variable (our preferred model), $G_0^2 = 12.309$ and the likelihood r-squared equals

$$\text{likelihood } r^2 = 1 - \exp\left(-\frac{12.309}{247}\right) = 0.05$$

Though this model is better than no model, there is still much variability left to explain in these data.

*Confidence interval for the slope*

A confidence interval associated with the slope of an explanatory variable can be computed using the Wald's normal approximation for large samples (equation 12.15). Ryan (1997, page 272) comments that the Wald's confidence intervals "should be used cautiously, especially if the sample size is not large".

$$\text{CI on } \beta: \quad \beta \pm z_{1-a/2} \bullet se_\beta \quad\quad (12.15)$$

For the TCE exceedance data a 95% confidence interval on the slope for Popden is

$$0.174 \pm 1.96 \bullet 0.05 = [0.077, 0.272]$$

*Odds ratios*

In the logistic regression equation (equation 12.9), the slope $b_1$ describes the change in the log of the odds ratio (logit) per unit change in $x_1$. Exponentiating, $\exp(b_1)$ is the amount by which the odds ratio changes for a unit change in $x_1$. For TCE exceedances, the odds ratio increases for a unit change in Popden by a multiplicative factor of $\exp(0.174) = 1.19$, where 0.174 is the estimated slope for Popden. For example, at a population density of 5 the probability of exceeding 5 µg/L TCE is 9%. The odds ratio is $\frac{0.09}{0.91}$, or 0.10. For the same conditions but at a population density of 6, the odds ratio will increase to 0.10 times 1.19, or 0.119. From equation 12.10, this equals a probability of exceedance $\pi$ of 0.119/1.119, or 10.7%.

Though $\exp(\beta)$ is the best point estimate for the rate of change of the odds ratio as a function of the explanatory variable, a 95% confidence interval for that change also can be calculated. If the Wald's estimates of standard deviation are acceptable, the confidence interval on the slope can be exponentiated into a confidence interval on the change in odds ratios:

CI on the change in the odds ratio $[\pi_L, \pi_H] = \exp\left[\beta \pm z_{1-a/2} \bullet se_\beta\right]$     (12.16)

For the TCE data, the 95% confidence interval on the rate of change of the odds ratio, as a function of a unit change in population density, is

$$\exp\left[\beta \pm z_{1-a/2} \bullet se_\beta\right] = \exp\left[0.174 - 1.96 \bullet 0.05, 0.174 + 1.96 \bullet 0.05\right] = \exp[0.077, 0.272]$$
$$= [1.08, 1.31]$$

With 95% confidence, the rate of change of the odds ratio is between 1.08 and 1.31 for a unit change in population density. This is the confidence interval printed out by Minitab® on the line for each explanatory variable in logistic regression.

*Confidence interval for the proportion p*

Of great interest is the estimated or predicted probability of detection $\pi$ as a function of reasonable values for the explanatory variables: the S-shaped logistic surface. This is provided by equation 12.10. Also of interest is a confidence interval around the fitted surface. How precisely is the estimate of $\pi$ known? An estimated probability of exceedance of 40%, or 0.4, may lead to different consequences if it is known with relative precision, say $0.4 \pm 0.05$, than if it can be estimated only crudely, say as $0.4 \pm 0.5$. In the latter case, the commitment of scarce resources or expensive solutions would probably not be justified.

First the point on the logistic regression surface $\pi$ is estimated. The tenth observation in the TCE data set has a population density of 11. The estimated probability of detection $\pi$ for this observation based on the model with Popden as the sole explanatory variable is:

$$\pi = \frac{\exp(-3.17 + 0.174 \bullet 11)}{1 + \exp(-3.17 + 0.174 \bullet 11)} = \frac{\exp(-1.256)}{1 + \exp(-1.256)} = 0.22$$

Overall, 22% of the samples collected for that population density are expected to have TCE concentrations greater than 5 µg/L.

An approximate $(1-\alpha)\%$ confidence interval around $\pi$, based on Wald's statistics, can be computed as equation 12.17 (Ryan, 1997):

$$[\pi_L, \pi_U] = \frac{\exp\left(b_0 + b_i X_i \pm z_{1-\alpha/2} s \sqrt{h_{ii}}\right)}{1 + \exp\left(b_0 + b_i X_i \pm z_{1-\alpha/2} s \sqrt{h_{ii}}\right)}$$     (12.17)

where $h_{ii}$ is the leverage statistic for the $i^{th}$ observation (a statistic that is large for outlying observations), and s is an estimate of the error around the estimated proportions. Ryan (1997) notes that this equation is only close to a true $(1-\alpha)\%$ interval when sample sizes are "very large". I would propose that "very large" be at least 100 observations total, with no fewer than 20 in either category. Fortunately, the TCE data meet this criteria.

The leverage statistic is a multivariate measure of the influence each individual observation has on the final regression model. Leverage measures by how much the estimates of slope and intercept would change when an observation is deleted from the data set. Outlying observations away from the main 'cloud' of data have higher leverage, and deleting those observations changes the estimates of slope and intercept more so than observations closer in to the bulk of the data. Minitab® will compute and save $h_{ii}$ as the leverage (Hi), if requested after clicking on the 'Storage' button. Other software packages do much the same.

However, the value for s is a matrix, and software programs do not print it out. To work around this, the standard error of the logit $s\sqrt{h_{ii}}$ , which changes based on the values of the explanatory variable(s), can be computed using the variance-covariance matrix. This matrix is output by logistic regression software. The standard error of the logit can be computed using equation 12.18 (based on Hosmer and Lemeshow, 2000, p. 19):

$$s\sqrt{h_{ii}} = \sqrt{Var(b_0) + x^2 Var(b_1) + 2xCov(b_0, b_1)} \qquad (12.18)$$

where the variance of the intercept $b_0$ and slope $b_1$ are the squares of their respective standard errors provided in the table of partial statistics. Equation 12.18 describes the standard error of the logit $SE\left[\ln\left(\dfrac{\pi}{1-\pi}\right)\right]$, or equivalently, the standard error of the linear estimates $SE\left[b_0 + b_j X_j\right]$. Variance estimates for each parameter, along with the covariance between all pairs of parameters $Cov(b_0, b_1)$ can be stored as an option within Minitab®.

For the TCE - Popden model, the variance-covariance matrix is stored as matrix M1 and printed:

```
Matrix XPWX5
  0.183850   -0.0188711
 -0.018871    0.0025017
```

The value 0.183850 is the variance of $b_0$, and equals the square of the standard error 0.428778 reported in the table of partial statistics. The value 0.0025017 is the variance of $b_1$, and equals the square of the standard error 0.0500166 reported in the table of partial statistics. The value −0.0188711 is the covariance between $b_0$ and $b_1$. The estimated standard error of the logit is therefore (following equation 12.18):

$$s\sqrt{h_{ii}} = \sqrt{0.183850 + x^2 \bullet 0.0025017 - 2x \bullet 0.0188711}$$

The probability of exceedance $\pi$ for a population density (x) of 11 was previously computed to be 0.22. A 95% confidence interval around this estimate is computed as:

$$s\sqrt{h_{ii}} = \sqrt{0.183850 + (11)^2 \bullet 0.0025017 - 22 \bullet 0.0188711} = \sqrt{0.0714} = 0.267$$

$$\pi = \frac{\exp(-3.17 + 0.174 \bullet 11 \pm 1.96 \bullet 0.267)}{1 + \exp(-3.17 + 0.174 \bullet 11 \pm 1.96 \bullet 0.267)} = \frac{\exp(-1.256 \pm 0.524)}{1 + \exp(-1.256 \pm 0.524)}$$

$$= \left[ \frac{0.1687}{1.1687}, \frac{0.4808}{1.4808} \right] = [0.144, 0.325].$$

At a population density of 11, a 95% confidence interval for the probability of TCE concentrations exceeding 5 µg/L is between 14.4 and 32.5 percent, with the best single estimate of the probability at 22 percent. This confidence interval will be accurate when there are good estimates of the variances and covariance of the parameters, resulting in a good approximation to the true standard error. This is more likely for large sample sizes, such as for the TCE data.

For multiple explanatory variables the process is similar to the above. There will be multiple covariance estimates between pairs of parameters. For a further discussion of how confidence intervals on $\pi$ might be computed using the variance-covariance matrix output by statistical software, see Chapter 2 of Hosmer and Lemeshow (2000).

As an alternative for computing confidence intervals with smaller samples, repeated logistic regressions can be performed to produce bootstrapped estimates for $\pi$. This will obviously be computer intensive, but the 100*$\alpha$/2th and 100*(1-$\alpha$/2)th percentiles of the bootstrapped estimates of $\pi$ will form the ends of an $\alpha$% confidence interval for the desired proportion. This bootstrapped interval does not require an assumption of normality for Wald's statistics that cannot be met by the smaller sample sizes common to most environmental studies.

### Additional methods for censored regression

Other methods can also be used for regression models with censored data. Three possibilities are: Cox proportional hazards models, Buckley-James regression, and quantile regression. The first two methods are regularly applied to censored data in the field of survival analysis. Proportional hazards is a standard tool for the biostatistician, and is discussed in essentially all textbooks on survival analysis. Buckley-James regression is a more specialized procedure found in some software and textbooks. Quantile regression is a newer method used for analysis of uncensored data, and appears in current statistical journals, but not yet in statistics textbooks.

Cox proportional hazards is a semiparametric method used to estimate the proportional effect of one or more explanatory variables $X_k$ on the censored Y variable, without modeling the underlying relationship between Y and the Xs. As such it does not result in a regression equation – there is no intercept and no functional form that will predict Y from the Xs. Instead, a rate increase is modeled, the change in the "hazard function" h(Y), as a function of the explanatory variables X:

$$h(Y) = h_0(Y) \bullet e^{\beta_i X_i} \tag{12.18}$$

A hazard function is the ratio of the probability density function f(Y) to the survival function S(Y). It is the "'approximate' probability of an individual of age x experiencing the event in the next instant" (Klein and Moeschberger, 2003). Translated into environmental applications, it is the approximate probability of a concentration falling below a detection limit as concentration decreases a minute amount. The concept of a hazard function has not translated well to environmental studies, perhaps leading to the lack of its use there.

However, the frequent use of proportional hazards models in medical studies may warrant a new look at these models for environmental applications. The focus of interest would be in the slope coefficient $\beta$. The $\beta$ slope coefficient in equation 12.18 is the source of statements reported to the general public in a form something like "the medication resulted in a 3-fold decrease in risk of heart attack". By keeping the exact form of the $h_0$ base function unspecified, no functional form of the relationship between X and Y need be assumed. The risk of heart attack for individuals may be high, or low, and may or may not be linear with respect to diet and exercise. None of that matters, because it is all contained in the unspecified base risk term $h_0$. What is of interest is that, given whatever risk there may be, that risk proportionally changes as X changes. All terms but the effect of X are "blocked out" by putting them into the base risk $h_0$.

Proportional hazards might be of use in environmental studies with censored data for performing procedures similar in purpose to analysis of covariance. With analysis of covariance, all effects on concentration other than the one to be studied are adjusted for. If X were a measure of exposure, such as industrial activity or proportion of soils with high metals content, the rate of risk for increasing the resultant concentration Y as X changes could be determined without assuming an underlying model for other effects that might be occurring. Proportional hazards could also be useful as a screening tool for determining which of several X variables has a significant effect on the distribution of concentrations Y, in a mode not unlike stepwise regression is now often (though perhaps inappropriately) used. The reader is referred to textbooks on survival analysis such as Lee and Wang (2003), Klein and Moeschberger (2003), Kalbfleisch and Prentice (2002), or Collett (2003) for a much more thorough treatment of hazard functions and proportional hazard models.

Buckley and James (1979) presented a method to construct a linear regression model for right-censored response variables (the x variables cannot be censored) without assumption of a normal distribution. The method is an iterative procedure to find a slope $\beta$ that fits a weighted combination of uncensored values for Y and the censored survival function for right-censored data. Ireson and Rao (1985) compared this method to the Theil-Sen slope, finding that Buckley-James regression always had larger confidence intervals than the Theil-Sen method, and so comparatively lacked precision. Buckley-James also suffers from a possible failure for the algorithm to converge to an answer. Wilcox (1998) showed that a modified form of Buckley-James regression had much larger bias than did Theil-Sen regression.

From these studies there appears little reason to prefer Buckley-James regression over the Theil-Sen methods described in this chapter.

Quantile regression (Cade and Noon, 2003) estimates the conditional quantiles or percentiles of a distribution as a linear function of explanatory variables. The impetus in ecology seems to be for fitting data with heterogeneous errors – residuals that show a pattern of increasing variance. For such data the slope of Y with X may be moderate for central data, such as near the mean or median, but larger slopes would better characterize the data when higher quantiles are of interest. The objective is something like "How do concentrations that are exceeded only 25% and 10% of the time change as a function of X?" Multiple equations for simultaneously fitting different quantiles are usually constructed.

Quantile regression has not yet been applied to censored data. Slope coefficients are normally calculated for uncensored data using weighted absolute deviation. These equations would have to be solved by maximum likelihood in order to correctly incorporate censored observations. In any case, quantile regression is not yet found in commercial statistical software, and routine application to censored data will require further study.

# Exercises

12-1    Atrazine concentrations were measured in streams across the Midwestern
        United States (Mueller et al., 1997).  Data are found in recon.xls.  Measured
        at each site were the following explanatory variables:

| Name | Description |
|------|-------------|
| Area | Basin size |
| Applic | Atrazine application rate, estimated from statewide estimates |
| Corn% | Percent of land area of watershed planted in corn |
| Soilgp | Soil hydrologic group, a measure of soil permeability found in STATSGO. |
| Temp | Annual average temperature  (a north – south indicator) |
| Precip | Annual average precipitation  (mostly an east – west indicator) |
| Dyplant | Days since planting (and therefore since last atrazine application) |
| Pctl | Percentile of streamflow (standardizes across streams of varying size) |
| Atraconc | Atrazine concentration, in ug/L |

Using censored parametric regression, build a multiple regression model to
relate atrazine concentrations to the variables in the list above.  Determine
what units are the best to be working in before settling on a final model.
Find the explanatory variables which are all significant at alpha = 0.05.

12-2    Brumbaugh et al. (2001) measured mercury concentrations in fish of
        approximately the same age and trophic level across the United States.
        Determine a regression equation for the dependence of mercury ("Hg") on
        one or more of the possible explanatory variables listed below.
        Transformation of the explanatory variables may be required.  All explana-
        tory variables included in the model should have a p-value of 0.05 or less.
        The data are found in HgFish.xls.

| Name | Description |
|------|-------------|
| WatMeHg | Methyl mercury concentrations in stream water |
| WatTotHg | Total mercury concentrations in stream water |
| SedMeHg | Methyl mercury concentrations in stream sediments |
| SedTotHg | Total mercury concentrations in stream sediments |
| WatDOC | Dissolved organic carbon concentrations in stream water |
| SedLOI | Loss of ignition (a measure of organic carbon content) in stream sediment |
| SedAVS | Sediment acid-volatile sulfides |
| % wetland | Percent of the basin occupied by wetlands |

**12-3**    Using the data in recon.xls collected by Mueller et al. (1997), compute a
logistic regression equation for predicting the probability of observing an
atrazine concentration above 1 μg/L. The variable GT_1 has a value equal
to 1 for all atrazine concentrations greater than 1 μg/L, and 0 otherwise.
Candidate explanatory variables are the same as those considered in
Exercise 12-1.

# APPENDIX

# DATASETS

The author's appreciation is extended to each of the scientists who provided their data for use in this book, especially to those who directly provided data not available in published reports. All data sets listed can be found at:

**http://www.practicalstats.com/nada**

in both Microsoft Excel (.xls) and Minitab® worksheet (.mtw) formats.

AsExample    Artificial numbers representing arsenic concentrations in a drinking water supply.
File Name:    AsExample.xls
Reference:    None. Generated.
Objective:    Determine what can be done with data where all values are below the reporting limit.
Censoring:    A detection limit at 1, and a reporting limit at 3 µg/L.
Used in:    Chapter 8

Atra    Atrazine concentrations in a series of Nebraska wells before (June) and after (September) the growing season.
File Name:    Atra.xls
Reference:    Junk et al., 1980, *Journal of Environmental Quality* 9, pp. 479–483.
Objective:    Determine if concentrations increase from June to September.
Censoring:    One detection limit, at 0.01 µg/L.
Used in:    Chapters 4, 5, and 9

AtraAlt    Atrazine concentrations altered from the Atra data set so that there are more nondetects, adding a second detection limit at 0.05.
File Name:    AtraAlt.xls
Reference:    Altered from the data of Junk et al., 1980, *Journal of Environmental Quality* 9, pp. 479–483.
Objective:    Determine if concentrations increase from June to September.
Censoring:    Two detection limits, at 0.01 and 0.05 µg/L.
Used in:    Chapters 5 and 9

**Atrazine**          The same atrazine concentrations as in Atra, stacked into one col-
                      umn (col.1).   Column 2 indicates the month of collection.
                      Column 3 indicates which data are below the detection limit –
                      those with a value of 1.
**File Name:**        Atrazine.xls
**Reference:**        Junk et al., 1980, *Journal of Environmental Quality* 9, pp.
                      479–483.
**Objective:**        Determine if concentrations increase from June to September.
**Censoring:**        One detection limit, at 0.01 µg/L.
**Used in:**          Chapter 9


**Bloodlead**         Lead concentrations in the blood of herons in Virginia.
**File Name:**        Bloodlead.xls
**Reference:**        Golden et al., 2003, *Environmental Toxicology and Chemistry* 22,
                      1517–1524.
**Objective:**        Compute interval estimates for lead concentrations.
**Censoring:**        One detection limit, at 0.02 µg/g.
**Used in:**          Chapter 7


**Cd**                Cadmium concentrations in fish for two regions of the Rocky
                      Mountains.
**File Name:**        Cd.xls
**Reference:**        none.  Data modeled after several reports.
**Objective:**        Determine if concentrations are the same or different in fish liv-
                      ers of the two regions.
**Censoring:**        Four detection limits, at 0.2, 0.3, 0.4, and 0.6 µg/L.
**Used in:**          Chapter 9


**ChlfmCA**           Chloroform concentrations in groundwaters of California.
**File Name:**        ChlfmCA.xls
**Reference:**        Squillace et al., 1999, *Environmental Science and Technology* 33,
                      4176–4187.
**Objective:**        Determine if concentrations differ between urban and rural areas.
**Censoring:**        Three detection limits, at 0.05, 0.1, and 0.2 µg/L.
**Used in:**          Chapter 9


**CuZn**              Copper and zinc concentrations in ground waters from two zones
                      in the San Joaquin Valley of California.  The zinc concentrations
                      were used.
**File Name:**        CuZn.xls

Reference:      Millard and Deverel, 1988, *Water Resources Research* 24, pp.
                2087–2098.
Objective:      Determine if zinc concentrations differ between the two zones.
Censoring:      Zinc has two detection limits, at 3 and 10 µg/L.
Used in:        Chapters 4, 5 and 9

## CuZnAlt

Zinc concentrations of the CuZn data set; concentrations in the
Alluvial Fan zone have been altered so that there are more nondetects. This produces a greater signal, even with more nondetects.

File Name:      CuZnAlt.xls
Reference:      Altered from the data of Millard and Deverel, 1988, *Water
                Resources Research* 24, pp. 2087–2098.
Objective:      Determine if zinc concentrations differ between the two zones.
Censoring:      Zinc has two detection limits, at 3 and 10 µg/L.
Used in:        Chapter 9

## DFe

Dissolved iron concentrations over several years in the Brazos
River, Texas. Summer concentrations were used.

File Name:      DFe.xls
Reference:      Hughes and Millard, 1988, *Water Resources Bulletin* 24, pp.
                521–531.
Objective:      Determine if there is a trend over time.
Censoring:      Iron has two detection limits, at 3 and 10 µg/L.
Used in:        Chapters 5, 11 and 12

## Doc

Dissolved organic carbon concentrations in ground waters of irrigated and non-irrigated areas.

File Name:      DOC.xls
Reference:      Junk et al., 1980, *Journal of Environmental Quality* 9, pp.
                479–483.
Objective:      Determine if concentrations differ between irrigated and non-irri-
                gated areas.
Censoring:      One detection limit at 0.2 µg/L.
Used in:        Chapter 9

## Golden

Lead concentrations in the blood and several organs of herons in
Virginia.

File Name:      Golden.xls
Reference:      Golden et al., 2003, *Environmental Toxicology and Chemistry* 22,

|                |                                                                                                                                                          |
|----------------|----------------------------------------------------------------------------------------------------------------------------------------------------------|
|                | pp. 1517–1524.                                                                                                                                            |
| Objective:     | Determine the relationships between lead concentrations in the blood and various organs. Do concentrations reflect environmental lead concentrations, as represented by dosing groups? |
| Censoring:     | One detection limit, at 0.02 µg/g.                                                                                                                        |
| Used in:       | Chapters 10 and 11                                                                                                                                        |

**Hatchery**    Proportions of detectable concentrations of antibiotics (µg/L) in drainage from fish hatcheries across the United States.

File Name:     Hatchery.xls

Reference:     Thurman et al., 2002, Occurrence of antibiotics in water from fish hatcheries. USGS Fact Sheet FS 120-02.

Objective:     Compute confidence intervals and tests on proportions.

Censoring:     One detection limit for each compound, all at 0.05 µg/L.

Used in:       Chapters 8 and 9

**HgFish**      Mercury concentrations in fish across the United States.

File Name:     HgFish.xls

Reference:     Brumbaugh et al., 2001, USGS Biological Science Report BSR-2001-0009.

Objective:     Do mercury concentrations differ by land use of the watershed? Can concentrations be related to water and sediment characteristics of the streams?

Censoring:     Three detection limits, at 0.03, 0.05, and 0.10 µg/g wet weight.

Used in:       Chapters 10, 11 and 12

**MDCu+**       Copper concentrations in ground water from the Alluvial Fan zone in the San Joaquin Valley of California. One observation was altered to become a <21, larger than all of the detected observations (the largest detected observation is a 20).

File Name:     MDCu+.xls

Reference:     Millard and Deverel, 1988, *Water Resources Research* 24, pp. 2087–2098.

Objective:     Calculation of summary statistics when the largest observation is censored.

Censoring:     Five detection limits, at 1, 2, 5, 10 and 20 µg/L. An additional artificial detection limit of 21 was added to illustrate a point.

Used in:       Chapter 6

Oahu             Arsenic concentrations (µg/L) in an urban stream, Manoa Stream
                 at Kanewai Field, on Oahu, Hawaii.
File Name:       OahuAs.xls
Reference:       Tomlinson, 2003, Effects of Ground-Water/Surface-Water
                 Interactions and Land Use on Water Quality. Written communi-
                 cation (draft USGS report).
Objective:       Characterize conditions by computing summary statistics.
Censoring:       Three detection limits, at 0.9, 1, and 2 µg/L. Uncensored values
                 reported below the lowest detection limit indicate that informative
                 censoring may have been used, and so the results are likely biased
                 high.
Used in:         Chapter 6


Recon            Atrazine concentrations in streams throughout the Midwestern
                 United States.
File Name:       Recon.xls
Reference:       Mueller et al., 1997, *Journal of Environmental Quality* 26,
                 1223–1230.
Objective:       Develop a regression of model for atrazine concentrations using
                 explanatory variables.
Censoring:       One detection limit, at 0.05 µg/L.
Used in:         Chapter 12


Roach            Lindane concentrations in fish from tributaries of the Thames
                 River, England.
File Name:       Roach.xls
Reference:       Yamaguchi et al., 2003, *Chemosphere* 50, 265–273.
Objective:       Determine whether lindane concentrations are the same at all
                 sites.
Censoring:       One detection limit at 0.08 µg/kg.
Used in:         Chapter 9


SedPb            Lead concentrations in stream sediments before and after wild-
                 fires.
File Name:       SedPb.xls
Reference:       Eppinger et al., 2003, USGS Open-File Report 03-152.
Objective:       Determine whether lead concentrations are the same pre- and
                 post-fire.
Censoring:       One detection limit at 4 µg/L.
Used in:         Chapter 9

**Silver**                 Silver concentrations in a standard solution sent to 56 laboratories
                           as part of a quality assurance program.
File Name:                 Silver.xls
Reference:                 Helsel and Cohn, 1988, *Water Resources Research* 24,
                           1997–2004.
Objective:                 Estimate summary statistics for the standard solution. The medi-
                           an or mean might be considered the 'most likely' estimate of the
                           concentration.
Censoring:                 Twelve detection limits, the largest at 25 µg/L.
Used in:                   Chapter 6

**Tbl1_1**                 Contaminant concentrations in test and a control group.
File Name:                 Tbl1_1.xls
Reference:                 None. Generated data.
Objective:                 Determine whether a test group has higher concentrations than a
                           control group.
Censoring:                 Three detection limits, at 1, 2, and 5 µg/L.
Used in:                   Chapter 1, Table 1.1

**TCE**                    TCE concentrations (µg/L) in ground waters of Long Island, New
                           York. Categorized by the dominant land use type (low, medium,
                           or high density residential) surrounding the wells.
File Name:                 TCE.xls
Reference:                 Eckhardt et al., 1989, USGS Water Resources Investigations
                           Report 86-4142.
Objective:                 Determine if concentrations are the same for the three land use
                           types.
Censoring:                 Four detection limits, at 1,2,4 and 5 µg/L.
Used in:                   Chapter 10

**TCEReg**                 TCE concentrations (µg/L) in ground waters of Long Island, New
                           York, along with several possible explanatory variables.
File Name:                 TCEReg.xls
Reference:                 Eckhardt et al., 1989, USGS Water Resources Investigations
                           Report 86-4142.
Objective:                 Determine if concentrations are related to one or more explanato-
                           ry variables.
Censoring:                 Four detection limits, at 1,2,4 and 5 µg/L. One column indicates
                           whether concentrations are above or below 5.
Used in:                   Chapter 12

**Thames**      Dieldrin, lindane and PCB concentrations in fish of the Thames River and tributaries, England.

File Name:    Thames.xls

Reference:    Yamaguchi et al., 2003, *Chemosphere* 50, 265–273.

Objective:    Determine if concentrations differ among sampling sites. Are dieldrin and lindane concentrations correlated?

Censoring:    One detection limit per compound.

Used in:    Chapters 11 and 12

# REFERENCES

Ahn, H., 1998, Estimating the mean and variance of censored phosphorus concentrations in Florida rainfall: *Journal of the American Water Resources Association* 34, 583–593.

Aitchison, J. and I.A.C. Brown, 1957, *The Lognormal Distribution.* Cambridge University Press, Cambridge, 176 pp.

Akritas, M.G., 1986, Bootstrapping the Kaplan-Meier estimator: *Journal of the American Statistical Association* 81, 1032–1038.

Akritas, M.G., 1992, Rank transform statistics with censored data: *Statistics and Probability Letters* 13, 209–221.

Akritas, M.G., 1994, Statistical analysis of censored environmental data: Chapter 7 of the *Handbook of Statistics*, Volume 12, edited by G.P. Patil and C. R. Rao. North-Holland, Amsterdam, 927 pp.

Akritas, M.G., S.A. Murphy and M.P. LaValley, 1995, The Theil-Sen estimator with doubly-censored data and applications to astronomy: *Journal of the American Statistical Association* 90, 170–177.

Akritas, M.G. and J. Siebert, 1996, A test for partial correlation with censored astronomical data: *Monthly Notices of the Royal Astronomical Society* 278, 919–924.

Allison, P.D., 1995, *Survival Analysis Using the SAS® System: A Practical Guide.* SAS Institute, Inc., Cary, NC, 292 pp.

ASTM, 1983. Sec. D4210, American Society of Testing Materials.

ASTM, 1991. Sec. D6091, Standard practice for 99%/95% interlaboratory detection estimate (IDE) for analytical methods with negligible calibration error. American Society of Testing Materials.

ASTM, 2000. Sec. D 6512, Standard practice for interlaboratory quantitation estimate (IQE), American Society of Testing Materials.

Beal, S.L., 2001, Ways to fit a PK model with some data below the quantification limit: *Journal of Pharmacokinetics and Pharmacodynamics* 28, 481–505.

Boos, D. D., and J. M. Hughes-Oliver, 2000, How large does n have to be for Z and t intervals?: *American Statistician* 54, 121–126.

Borgan, O. and K.A. Liestøl, 1990. A note on the confidence intervals and bands for the survival curve based on transformations: *Scandinavian Journal of Statistics* 17, 35–41.

Brookmeyer, R. and J. Crowley, 1982, A confidence interval for the median survival time: *Biometrics* 38, 29–41.

Brumbaugh, W.G., D.P. Krabbenhoft, D.R. Helsel, J.G. Wiener, and K.R. Echols, 2001, A national pilot study of mercury contamination of aquatic ecosystems along multiple gradients – bioaccumulation in fish: U.S. Geological Survey Biological Science Report BSR-2001-0009, 25 pp.

Buckley, J, and I. James, 1979, Linear regression with censored data: *Biometrika* 66, 429–436.

Buckley, T.J., J. Liddle, D. L. Ashley, D. C. Paschal, V. W. Burse, L.L. Needham, and G. Akland, 1997, Environmental and biomarker measurements in nine homes in the lower Rio Grande valley: Multimedia results for pesticides, metals, PAHs, and VOCs; *Environmental Pollution* 23, 705–722.

Cade, B.S. and B.R. Noon, 2003, A gentle introduction to quantile regression for ecologists: *Ecological Environment* 1, 412–420.

Chay, K.Y., and B.E. Honore, 1998, Estimation of censored semiparametric regression models: An application to changes in Black-White earnings inequality during the 1960s: *Journal of Human Resources* 33, 4–38.

Clarke, J. U., 1998, Estimation of censored data methods to allow statistical comparisons among very small samples with below detection limit observations: *Environmental Science and Technology* 32, 177–183.

Cohen, A. C., 1959, Simplified estimators for the normal distribution when samples are singly censored or truncated: *Technometrics* 1, 217–237.

Cohen, A. C., 1961, Tables for maximum likelihood estimates: Singly truncated and singly censored samples. *Technometrics* 3, 535–541.

Cohn, T.A., 1988, Adjusted maximum likelihood estimation of the moments of lognormal populations from type I censored samples: U.S. Geological Survey Open-File Report 88-350, 34 pp.

Collett, D., 2003, *Modeling Survival Data in Medical Research*, Second edition. Chapman and Hall/CRC, London, 391 pp.

Conover, W.J. and R.L. Iman, 1981, Rank transformations as a bridge between parametric and nonparametric statistics: *American Statistician* 35, 124–129.

Conover, W.J., 1968, Two k-sample slippage tests: *Journal of the American Statistical Association* 63, 614–626.

Conover, W.J., 1999, *Practical Nonparametric Statistics*, Third Edition. Wiley, New York, 584 pp.

Currie, L. A., 1968, Limits for qualitative detection and quantitative determination: *Analytical Chemistry* 48, 586–593.

Dietz, E.J., 1987, A comparison of robust estimation in simple linear regression: *Communications in Statistical Simulation* 16, 1209–1227.

Dietz, E.J., 1989, Teaching regression in a nonparametric statistics course: *American Statistician* 43, 35–40.

Dietz, E.J. and T.J. Killeen, 1981, A nonparametric multivariate test for monotone trend with pharmaceutical applications: *Journal of the American Statistical Association* 76, 169–174.

Eckhardt, D.A., W.J. Flipse and E.T. Oaksford, 1989, Relation between land use and ground-water quality in the upper glacial aquifer in Nassau and Suffolk Counties, Long Island NY: U.S. Geological Survey Water Resources Investigations Report 86-4142, 26 pp.

Efron, B., 1981, Censored data and the bootstrap: *Journal of the American Statistical Association* 374, 312–319.

Efron, B. and R. J. Tibshirani, 1986, Bootstrap methods for standard errors, confidence intervals and other measures of statistical accuracy: *Statistical Science* 1, 54–77.

El-Shaarawi, A.H. and S.R. Esterby, 1992, Replacement of censored observations by a constant: An evaluation: *Water Research* 26, 835–844.

Emerson, J.D., 1982, Nonparametric confidence intervals for the median in the presence of right censoring: *Biometrics* 38, 17–27.

Eppinger, R.G., P.H. Briggs, B. Rieffenberger, C. Van Dorn, Z.A. Brown, J.G. Crock, P.H. Hagemann, A. Meier, S.J. Sutley, P.M. Theodorakos, and S. A. Wilson, 2003, Geochemical data for stream sediment and surface water samples from Panther Creek, the middle fork of the Salmon River, and the main Salmon River, collected before and after the Clear Creek, Little Pistol, and Shellrock wildfires of 2000 in Central Idaho: U.S. Geological Survey Open-File Report 03-152. Available on CD.

Fong, D. Y. T., C.W. Kwan, K. F. Lam, and K. S. L. Lam (2003), Use of the sign test for the median in the presence of ties: *American Statistician* 57, 237–240.

Gehan, E. A., 1965, A generalized Wilcoxon test for comparing arbitrarily singly censored samples: *Biometrika* 52, 203–223.

Gibbons, R. D., 1995, Some statistical and conceptual issues in the detection of low level environmental pollutants: *Environmental and Ecological Statistics* 2, 125–145.

Gibbons, R. D., D. E. Coleman, and R. F. Maddalone, 1997, An alternative minimum level definition for analytical quantification: *Environmental Science and Technology* 31, 2071–2077.

Gibbons, R. D., and D. E. Coleman, 2001, *Statistical Methods for Detection and Quantification of Environmental Contamination.* Wiley, New York, 384 pp.

Gilbert, R.O. , 1987, *Statistical Methods for Environmental Pollution Monitoring.* Wiley, New York, 320 pp.

Gilbert, R.O. and R.R. Kinnison, 1981, Statistical methods for estimating the mean and variance from radionuclide data sets containing negative, unreported or less-than values. *Health Physics* 40, 377–390.

Gilliom, R. J., R. M. Hirsch, and E. J. Gilroy, 1984, Effect of censoring trace-level water-quality data on trend-detection capability: *Environmental Science and Technology* 18, 530–535.

Gilliom, R. J., and D. R. Helsel, 1986, Estimation of distributional parameters for censored trace level water quality data, 1. Estimation techniques: *Water Resources Research* 22, 135–146.

Gleit, A., 1985, Estimation for small normal data sets with detection limits. *Environmental Science and Technology* 19, 1201–1206.

Golden, N. H., B. A. Rattner, J. B. Cohen, D. J. Hoffman, E. Russek-Cohen, and M. A. Ottinger, 2003, Lead accumulation in feathers of nestling black-crowned night herons (Nycticorax nycticorax) experimentally treated in the field: *Environmental Toxicology and Chemistry* 22, 1517–1524.

Haas, C.N., and P.A. Scheff, 1990, Estimated of averages in truncated samples. *Environmental Science and Technology* 24, 912–919.

Hahn, G. J. and Meeker, W. Q., 1991, *Statistical Intervals: A Guide for Practitioners.* Wiley, New York, 392 pp.

Harris, M.L., J.E. Elliott, R.W. Butler, and L.K. Wilson, 2003, Reproductive success and chlorinated hydrocarbon contamination of resident great blue herons (Ardea herodias) from coastal British Columbia, Canada, 1977 to 2000: *Environmental Pollution* 121, 207–227.

Helsel, D.R. and T. A. Cohn, 1988, Estimation of descriptive statistics for multiply censored water quality data: *Water Resources Research* 24, 1997–2004.

Helsel, D.R. 1990, Less than obvious: Statistical treatment of data below the detection limit: *Environmental Science and Technology* 24, pp.1766–1774.

Helsel, D.R. and Gilliom, R.J., 1986, Estimation of distributional parameters for censored trace level water quality data, 2. Verification and applications: *Water Resources Research* 22, 147–155.

Helsel, D. R. and Hirsch, R. M., 2002, Statistical Methods in Water Resources. U.S. Geological Survey Techniques of Water Resources Investigations, Book 4, Chapter A3, 512 pp. Available at **http://water.usgs.gov/pubs/twri/twri4a3/**

Hettsmansperger, T.P., J.W. McKean, and S.J. Sheather, 1997, Rank-based analyses of linear models. Chapter 7 of the *Handbook of Statistics*, Volume 15, edited by G.S. Maddala and C. R. Rao. North-Holland, Amsterdam, 716 pp.

Hirsch, R.M. and J.R. Stedinger, 1987, Plotting positions for historical floods and their precision: *Water Resources Research* 23, 715–727.

Hirsch, R.M. and J.R. Slack, 1984, A nonparametric trend test for seasonal data with serial dependence: *Water Resources Research* 20, 727–732.

Hobbs, K. E., D. C. G. Muir, E. W. Bornb, R. Dietzc, T. Haugd, T. Metcalfee, C. Metcalfee and N. Øien, 2003, Levels and patterns of persistent organochlorines in mink whale (Balaenoptera acutorostrata) stocks from the North Atlantic and European Arctic; *Environmental Pollution* 121, 239–252.

Hollander, M. and D. A. Wolfe, 1999, *Nonparametric Statistical Methods*, Second Edition. Wiley, New York, 787 pp.

Hornung, R.W. and L.D. Reed, 1990, Estimation of average concentration in the presence of nondetectable values: *Applied Occupational Environmental Hygiene* 5, 46–51.

Hosmer, D.W. and S. Lemeshow, 2000, *Applied Logistic Regression*, Second Edition. Wiley, New York, 375 pp.

Hughes, J. P and Millard, S. P., 1988, A tau-like test for trend in the presence of multiple censoring points: *Water Resources Bulletin* 24, 521–531.

Huybrechts, T., O. Thas, J. Dewulf, and H. Van Langenhove, 2002, How to estimate moments and quantiles of environmental data sets with non-detected observations? A case study on volatile organic compounds in marine water samples: *Journal of Chromatography* 975, 123–133.

Ireson, M.J, and P.V. Rao, 1985, Interval estimation of slope with right-censored data: *Biometrika* 72, 601–608.

Isobe, T., E. D. Feigelson, and P. I. Nelson, 1986, Statistical methods for astronomical data with upper limits. II. Correlation and regression: *Astrophysical Journal* 306, 490–507.

Jeng, S.L. and W.Q. Meeker, 2001, Parametric simultaneous confidence bands for cumulative distributions from censored data: *Technometrics* 43, 450–461.

Junk, G. A., R.F. Spalding, and J.J. Richard, 1980, Areal, vertical, and temporal differences in ground-water chemistry: II. Organic constituents: *Journal of Environmental Quality* 9, 479–483.

Kahn, H.D., W.A. Telliard, and C.E. White, 1998, Comment on "An alternative minimum level definition for analytical quantification": *Environmental Science and Technology* 32, 2346–2348.

Kalbfleisch, J.D. and R. L. Prentice, 2002. *The Statistical Analysis of Failure Time Data*, Second edition. Wiley, New York, 439 pp.

Keith, L.H., 1992, *Environmental Sampling and Analysis: A Practical Guide*. Lewis Publishers, Chelsea, Michigan, 143 pp.

Kendall, M. G., 1955. *Rank Correlation Methods*, Second Edition. Charles Griffin and Company, London, 196 p.

Klein, J.P. and M.L. Moeschberger, 2003, *Survival Analysis: Techniques for Censored and Truncated Data*, Second edition. Springer, New York, 536 pp.

Kolpin, D. W., E. T. Furlong, M. T. Meyer, E. M. Thurman, S. D. Zaugg, L. Barber, and H. T. Buxton, 2002a, Pharmaceuticals, hormones, and other organic wastewater contaminants in U. S. streams, 1999–2000: A national reconnaissance; *Environmental Science and Technology* 36, 1202–1211.

Kolpin, D.W., J. E. Barbash, and R. J. Gilliom, 2002b, Atrazine and metolachlor occurrence in shallow ground water of the United States, 1993 to 1995: Relations to explanatory factors; *Journal of the American Water Resources Association* 38, 301–311.

Kroll, C.N. and J.R. Stedinger, 1996, Estimation of moments and quantiles using censored data: *Water Resources Research* 32, 1005–1012.

Land, C. E., 1972, An evaluation of approximate confidence interval estimation methods for lognormal means: *Technometrics* 14, 145–158.

Latta, R. B., 1981, A Monte Carlo study of some two-sample rank tests with censored data: *Journal of the American Statistical Association* 76, 713–719.

Lee, E. T., and J.W. Wang, 2003, *Statistical Methods for Survival Data Analysis*, Third edition. Wiley, New York, 534 pp.

Looney, S. W. and T. R. Gulledge, 1985, Use of the correlation coefficient with normal probability plots: *American Statistician* 39, 75–79.

Lundgren, R. F. and T. J. Lopes, 1999, Occurrence, distribution, and trends of volatile organic compounds in the Ohio River and its major tributaries, 1987–96; U.S. Geological Survey Water-Resources Investigations Report 99-4257, 89 pp.

Lynn, H.S., 2001, Maximum likelihood inference for left-censored HIV RNA data: *Statistics in Medicine* 20, 33–45.

Meeker, W.O. and L.A. Escobar, 1998, *Statistical Methods for Reliability Data*. Wiley, New York, 680 pp.

Miesch, A., 1967, Methods of computation for estimating geochemical abundance: U.S. Geological Survey Professional Paper 574-B, 15 pp.

Millard, S.P. and S. J. Deverel, 1988, Nonparametric statistical methods for comparing two sites based on data with multiple nondetect limits: *Water Resources Research* 24, 2087–2098.

Millard, S.P. and N. K. Neerchal, 2001, *Environmental Statistics with S-Plus*. CRC Press, New York. 830 pp.

Mueller, D.K., B.C. Ruddy, and W.A. Battaglin, 1997, Logistic model of nitrate in streams of the Upper Midwestern United States: *Journal of Environmental Quality* 26, 1223–1230.

Murphy, S.A., 1995, Likelihood ratio-based confidence intervals in survival analysis: *Journal of the American Statistical Association* 90, 1399–1405.

Nair, V.N., 1984, Confidence bands for survival functions with censored data: A comparative study: *Technometrics* 26, 265–275.

Nehls, G.J. and G. G. Akland, 1973, Procedures for handling aerometric data; *Journal of the Air Pollution Control Association* 23, 180–184.

Oakes, D., 1982, A concordance test for independence in the presence of censoring: *Biometrics* 38, 451–455.

Oblinger Childress, C.J., W. T. Foreman, B. F. Connor, and T. J. Maloney, 1999, New reporting procedures based on long-term method detection levels and some considerations for interpretations of water-quality data provided by the U.S. Geological Survey National Water Quality Laboratory: USGS Open-File Report 99–193, 19 pp.

O'Brien, P. C. and T. R. Fleming, 1987, A paired Prentice-Wilcoxon test for censored paired data: *Biometrics* 43, 169–180.

Owen, W., and T. DeRouen, 1980, Estimation of the mean for lognormal data containing zeros and left-censored values, with applications to the measurement of worker exposure to air contaminants: *Biometrics* 36, 707–719.

Perkins, J.L., G.N. Cutter, and M.S. Cleveland, 1990, Estimating the mean, variance, and confidence limits from censored (<limit of detection), lognormally-distributed exposure data: *American Industrial Hygiene Association Journal* 51, 416–419.

Peto, R. and J. Peto, 1972, Asymptotically efficient rank invariant test procedures (with discussion): *Journal of the Royal Statistical Society*, Series A 135, 185–206.

Porter, P. S., R. C. Ward, and H. F. Bell, 1988, The detection limit: *Environmental Science and Technology* 22, 856–861.

Prentice, R.L., 1978, Linear rank tests with right-censored data: *Biometrika* 65, 167–179.

Prentice, R.L. and P. Marek, 1979. A qualitative discrepancy between censored data rank tests: *Biometrics* 35, 861–867.

Rao, S.T., J.Y. Ku, and K.S. Rao, 1991. Analysis of toxic air contaminant data containing concentrations below the limit of detection: *Journal of the Air and Waste Management Association* 41, 442–448.

Rigo, H.G., 1999, Comment on "An alternate minimum level definition for analytical quantification": *Environmental Science and Technology* 33, 1311–1312.

Ryan, T.P., 1997, *Modern Regression Methods*. Wiley, New York, 515 pp.

Sanford, R.F., C.T. Pierson, and R.A. Crovelli, 1993. An objective replacement method for censored geochemical data: *Mathematical Geology* 25, 59–80.

Sen, P.K., 1968, Estimates of the regression coefficient based on Kendall's tau: *Journal of the American Statistical Association* 63, 1379–1389.

Shapiro, S.S. and M. B. Wilk, 1965. An analysis of variance test for normality (complete samples): *Biometrika* 52, 591–611.

She, N., 1997, Analyzing censored water quality data using a non-parametric approach: *Journal of the American Water Resources Association* 33, 615–624.

Shumway, R. H., A. S. Azari, and P. Johnson, 1989, Estimating mean concentrations under transformation for environmental data with detection limits: *Technometrics* 31, 347–356.

Shumway, R. H., R. S. Azari, and M. Kayhanian, 2002, Statistical approaches to estimating mean water quality concentrations with detection limits: *Environmental Science and Technology* 36, 3345–3353.

Simon, R. and Y.J. Lee, 1982, Nonparametric confidence limits for survival probabilities and median survival time: *Cancer Treatment Reports* 66, 37–42.

Singh, A. and J. Nocerino, 2002, Robust estimation of mean and variance using environmental data sets with below detection limit observations; Chemometrics and Intelligent Laboratory Systems 60, 69–86.

Singh, A K., A. Singh, and M. Engelhardt, 1997, The lognormal distribution in environmental applications: U.S. Environmental Protection Agency report EPA/600/R-97/006, 19 pp.

Slyman, D.J., A. de Peyster, and R.R. Donohoe, 1994, Hypothesis testing with values below detection limit in environmental studies: *Environmental Science and Technology* 28, 898–902.

Smith, D.E. and K.C. Burns, 1998, Estimating percentiles from composite environmental samples when all observations are nondetectable: *Environmental and Ecological Statistics* 5, 227–243.

Squillace, P.J., M.J. Moran, W.W. Lapham, C.V. Price, R.M. Clawges, and J.S. Zogorski, 1999, Volatile organic compounds in untreated ambient groundwater of the United States, 1985–1995: *Environmental Science and Technology* 33, 4176–4187.

Stephens, M.A., 1974. EDF statistics for goodness of fit and some comparisons: *Journal of the American Statistical Association* 69, 730–737.

Theil, H., 1950, A rank-invariant method of linear and polynomial regression analysis: *Nederl. Akad. Wetensch, Proceed.*, 53, 386–392.

Thompson, M.L. and K. P. Nelson, 2003, Linear regression with Type I interval- and left-censored response data: *Environmental and Ecological Statistics* 10, 221–230.

Thurman, E.M., J.E. Dietze, and E.A. Scribner, 2002, Occurrence of antibiotics in water from fish hatcheries. USGS Fact Sheet FS 120-02, 4 p. Available at: **http://pubs.water.usgs.gov/fs-120-02/**

Till, A. E., 2003, Comment on "Pharmaceuticals, hormones, and other organic wastewater contaminants in U.S. streams, 1999–2000": *Environmental Science and Technology* 37, 1052–1053.

Tobin, J., 1958, Estimation of relationships for limited dependent variables: *Econometrica* 26, 24–26.

Tomlinson, M.S., 2003, Effects of Ground-Water/Surface-Water Interactions and Land Use on Water Quality. Written communication in advance of becoming a USGS report.

Travis, C.C. and M.L. Land, 1990, Estimating the mean of data sets with nondetectable values. *Environmental Science and Technology* 24, 961–962.

U.S. Army Corps of Engineers, 1998, Evaluation of dredged material proposed for discharge in waters of the U.S. – Testing manual: published by USEPA, Office of Water, as EPA-823-B-98-004, see Appendix D, 113 pp.

U.S. Environmental Protection Agency, 1982, Definition and procedure for the determination of the method detection limit — revision 1.11; Code of Federal Regulations 40, Part 136, Appendix B., p. 565–567.

U.S. Environmental Protection Agency, 1998, Guidance for data quality assessment. Practical methods for data analysis; EPA/600/R-96/084. Available at: **http://www.epa.gov/swerust1/cat/epaqag9.pdf**

U.S. Environmental Protection Agency, 2002a, Development document for proposed effluent limitations guidelines and standards for the concentrated aquatic animal production industry point source category; EPA-821-R-02-016. Available at: **http://www.epa.gov/waterscience/guide/aquaculture/tdd/complete.pdf**

U.S. Environmental Protection Agency, 2002b, Guidance for comparing background and chemical concentrations in soils for CERCLA sites; EPA-540-R-01-003. Available at: **http://www.epa.gov/superfund/programs/risk/background.pdf**

U.S. Environmental Protection Agency, 2003, Technical support document for the assessment of detection and quantitation approaches: EPA-821-R-03-005. 71 pages. Available at: **http://www.epa.gov/waterscience/methods/det/dgch1-3.pdf**

Vance, D.E., Ehmann, W.D., and W.R. Markesbery, 1988, Trace element content in fingernails and hair of a nonindustrialized U.S. control population: *Biological Trace Element Research* 17, 109–121.

Velleman, P. F. and D. C. Hoaglin, 1981, *Applications, Basics, and Computing of Exploratory Data Analysis*: Duxbury Press, Boston, 354 pp.

Ware, J.H. and D.L. DeMets, 1976, Reanalysis of some baboon descent data. *Biometrics* 32, 459–463.

Wen, X.H., 1994, Estimation of statistical parameters for censored lognormal hydraulic conductivity measurements: *Mathematical Geology* 26, 717–731.

Weston, S.A. and W.Q. Meeker, 1991, Coverage probabilities of nonparametric simultaneous confidence bands for a survival function: *Journal of Statistical Computing Simulation*, 38, 83–97.

White, C. E. and H. D. Kahn, 1995, Discussion on the paper by R. D. Gibbons, "Some statistical and conceptual issues in the detection of low level environmental pollutants": *Environmental and Ecological Statistics* 2, 149–154.

Wilcox, R.R., 1998, Simulations on the Theil-Sen regression estimator with right-censored data: *Statistics and Probability Letters* 39, 43–47.

Yamaguchi, N., D. Gazzard, G. Scholey, and D.W. MacDonald, 2003, Concentrations and hazard assessment of PCBs, organochlorine pesticides and mercury in fish species from the upper Thames - River pollution and its potential effects on top predators: *Chemospere* 50, 265–273.

Zar, J. H., 1999, Biostatistical Analysis, Fourth edition. Prentice-Hall, Upper Saddle River, New Jersey, 875 pp.

# INDEX

accelerated failure time models, 201
air quality, 9, 13, 55, 56, 76
Akritas test, 158, 160
Akritas-Theil-Sen slope. See Theil-
Sen slope.
all nondetects, 119, 121, 127, 128
analysis of variance, 165, 183
ANOVA, 169
arbitrary censoring, 60, 168, 202, 204
astronomy, 9, 196, 237
balanced errors, 28
between the limits, 29, 61
bias, 29, 31, 77, 201
binomial probability, 103, 104, 105,
106, 112, 119, 121, 129
binomial test. See quantile test.
Bonferroni adjustment, 175, 176, 181
bootstrapping, 76, 84, 85, 96, 109,
113
boxplot, 43, 44, 131
Buckley-James regression, 210, 224,
225
calibration-based limits, 34
cdf, 47. See cumulative distribution
function.
censored data, 5, 9, 11, 14, 17, 19, 38,
242
censored regression, 18, 19, 134, 135,
138, 167, 169, 171, 204
Central Limit Theorem, 84, 86
Challenger accident, 1, 3
coefficient of determination, 187
Cohen's method, 56, 57, 58, 60, 74,
76, 186
compliance. See legal standard.
confidence bound,81, 84, 87, 88, 89,
97, 98, 99, 112
confidence interval, 81, 83, 84, 96,
154

confidence interval for percenitles,
112
confidence interval for the median,
98, 104, 109, 111
confidence intervals for percentiles,
98
constant standard deviation, 21
contingency table test, 167, 176, 177,
183
correlation, 19, 185
correlation coefficient, 211
coverage, 91
critical value, 22
cumulative distribution function, 14,
15, 47, 64, 124, 128, 143, 203
database, 37
decision level, 22
decision limit, 22
deleting nondetects, 2, 11, 43
detection limit, 5, 6, 12, 14, 21, 22,
23, 26
doubly-censored data, 210, 212, 237
edf, 47, 49, 53, 146. See empirical
distribution function.
empirical distribution function, 43, 47,
65
exceedance probabilities, 71
factors, 169, 171, 183
failure time models, 201
false negative, 25, 26. See Type II
error.
false positive, 23, 25. See Type I
error.
flipping data, 17, 50, 60, 63, 64, 65,
106, 107, 136, 146, 147, 148, 149,
150, 151, 152, 153, 159, 160, 176,
179, 183, 202
Gehan test, 143, 146, 150
generalized $r^2$, 187

## STATISTICS IN PRACTICE

*Human and Biological Sciences*

Brown and Prescott - Applied Mixed Models in Medicine
Ellenberg, Fleming and DeMets – Data Monitoring Committees in Clinical Trials: A Practical Perspective
Lawson, Browne and Vidal Rodeiro – Disease Mapping with WinBUGS and MLwiN
Lui – Statistical Estimation of Epidemiological Risk
Marubini and Valsecchi - Analysing Survival Data from Clinical Trials and Observation Studies
Parmigiani – Modeling in Medical Decision Making: A Bayesian Approach
Senn - Cross-over Trials in Clinical Research, Second Edition
Senn - Statistical Issues in Drug Development
Spiegelhalter, Abrams and Myles – Bayesian Approaches to Clinical Trials and Health-Care Evaluation
Whitehead - Design and Analysis of Sequential Clinical Trials, Revised Second Edition
Whitehead – Meta-Analysis of Controlled Clinical Trials

*Earth and Environmental Sciences*

Buck, Cavanagh and Litton – Bayesian Approach to Interpreting Archaeological Data
Glasbey and Horgan – Image Analysis in the Biological Sciences
Helsel – Nondetects And Data Analysis: Statistics for Censored Environmental Data
Webster and Oliver – Geostatistics for Environmental Scientists

*Industry, Commerce and Finance*

Aitken - Statistics and the Evaluation of Evidence for Forensic Scientists, Second Edition
Lehtonen and Pahkinen - Practical Methods for Design and Analysis of Complex Surveys, Second Edition
Ohser and Mücklich - Statistical Analysis of Microstructures in Materials Science